Introducing
Oceanography

Companion titles

Introducing Geology ~ A Guide to the World of Rocks (Second Edition 2010)

Introducing Palaeontology ~ A Guide to Ancient Life (2010)

Introducing Volcanology ~ A Guide to Hot Rocks (2011)

Introducing Geomorphology ~ A Guide to Landforms and Processes (2012)

Introducing Tectonics, Rock Structures and Mountain Belts (2012)

Introducing Meteorology ~ A Guide to Weather (Forthcoming 2012)

For further details of these and other Dunedin Earth and Environmental Sciences titles see:
www.dunedinacademicpress.co.uk

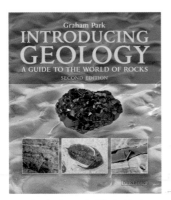

ISBN 978-1-906716-21-9

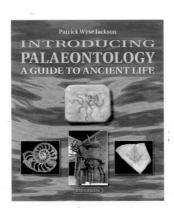

ISBN 978-1-906716-15-8

ISBN 978-1-906716-22-6

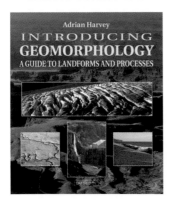

ISBN 978-19067163-25-7

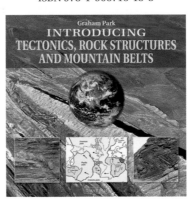

ISBN 978-1-906716-26-4

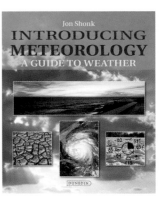

ISBN 978-1-780460-02-4

Introducing Oceanography

David N. Thomas and David G. Bowers

DUNEDIN

EDINBURGH ◆ LONDON

For several years we both had the privilege of teaching a course on
Shelf Sea Oceanography with Dr Sarah Jones (1962–2008).
Sarah was an inspiration to the students as well as to us.
We dedicate this book to her memory.

Published by
Dunedin Academic Press Ltd
www.dunedinacademicpress.co.uk

Head Office
Hudson House, 8 Albany Street
Edinburgh, EH1 3QB

London Office
The Towers, 54 Vartry Road,
London, N15 6PU

ISBN 9781780460017

British Library Cataloguing in Publication data
A catalogue record for this book is available from the British Library

Design and layout by Makar Publishing Production, Edinburgh
Printed and bound in Poland by Hussar Books

Contents

Preface

The study of the oceans, or oceanography, is of critical importance to all humans on the planet. Of the 198 nations in the world there are only 48 that are totally land-locked. But even those living in regions cut off from the oceans depend on shipping to transport goods and services. The oceans are a rich source of food for billions of humans. There is nowhere on the surface of the planet that does not rely on the effects that the oceans have on climate, and even more importantly in driving major water, oxygen, carbon dioxide, nitrogen and other cycles of the elements fundamental to life on Earth.

There is no escaping the importance of studying the oceans and understanding how they work. For oceanographers it is a privilege to work in such an important and exciting research field. As can be seen in Figure i.i, processes in the oceans span space and time scales from fractions of millimetres and seconds to tens of thousands of years, influencing millions of square kilometres.

No individual can study all aspects of oceanography by themselves. The oceans are complex chemical mixtures; the biological diversity and abundances are huge; a myriad of physical processes are vital for determining where the water goes and why. The oceans span the climate regions from the poles to the tropics. Oceanographers include specialists in physics, chemistry, biology, geology, mathematical modelling, atmospheric scientists, and space scientists working with satellites, among others. People working on the possibilities of life on extraterrestrial planets also look to the ocean for proxies of life further afield. To make any sense of their individual measurements it is often (mostly) vital that scientists from these different disciplines collaborate to synthesise their findings to understand the whole picture.

Although state-of-the-art technologies enable us to observe the oceans from space using satellites, there is only really one way to study oceanography, and that is to take to the seas in a research vessel. Such oceanographic cruises are voyages of discovery and thrilling opportunities. They are hard work and can last many months (although can be just a matter of hours or days). They can take the ships to regions of the world where wind and waves conspire to make life on board a ship very uncomfortable indeed. On the other hand, watching the sun set (or rise) on the horizon of a calm tropical ocean is quite the opposite.

In this short introduction we aim to give you a flavour of what oceanography is about. Naturally we can only touch on a few issues, but we hope to give you enough to make you want to find out more.

Note: all terms highlighted in **bold** are defined in the Glossary at the end of the book.

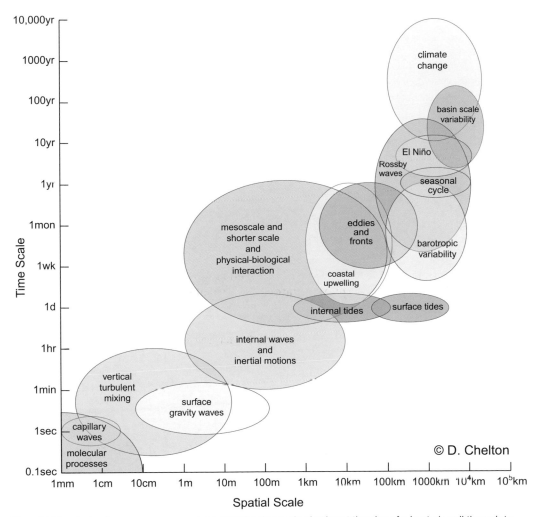

Figure i.i The study of ocean processes varies from the tiny scale physics at the size of a bacteria cell through to global ocean circulation patterns over tens of thousands of years.

1   The water of the oceans

Oceanography is the scientific study of the oceans, and of the organisms that live in them. Let's start by considering the dimensions of the subject before us. The oceans cover a large part (roughly 70%) of the earth's surface, but their vertical scale is small compared to the horizontal. Figure 1.1 shows a map of ocean depths. There are mountains and valleys on the sea floor, just as there are on dry land, but the average depth of all the oceans is about 4 km. This compares to the width of the oceans, which is variable, but which is of order 10,000 km. The ratio of the width to depth, the **aspect ratio**, is therefore 10,000:4, or 2,500:1. That's about the same as the aspect ratio of the page you are reading right now. Although oceans are thin compared to their width, changes with depth are very important, as we shall see in this chapter. For example, the temperature of ocean water can change as much in going down a few kilometres from the surface at the equator to the ocean floor as it can in travelling along the surface from the equator to the poles.

Around the margins of the oceans there are shallow water bodies called shelf seas. These are much smaller in horizontal extent (say 250 km) and shallower (typically 100 m) than the oceans, but they have a similar aspect ratio. Examples of shelf seas are the North Sea in Europe or the Yellow Sea in Asia.

1.1 Salinity and temperature of ocean water

The most obvious difference between ocean water and water that comes out of the tap is that the ocean is *salty*. It is not obvious where the salt comes from: rivers flowing into the ocean appear to carry perfectly fresh, drinkable water. In fact, river water contains trace amounts of ions produced by the weathering of rocks, but which are present in such low concentrations that we cannot taste them. Rivers are the largest source of the major ions in the oceans (Chapter 9), and these ions have accumulated over millions of years (it is estimated that sodium (Na^+) has a residence time of 78 million years and calcium (Ca^{2+}) around 1.1 million years in the ocean). A further source of ions is underwater volcanic eruptions. The fact that ocean salinity is not increasing perceptibly means that the ions introduced to the ocean are removed at about the same rate. Ions are removed by being incorporated into the biology and sediments, transferred to the atmosphere when bubbles burst on the ocean surface producing aerosols, evaporating and precipitating out of solution.

The average salt content of all the oceans is about 35 grams of salt per kilogram of water, or 35 parts per thousand, by mass. Since 1978, scientists have used practical salinity units (PSU) to express **salinity**. This is a ratio, and so has no units. In most cases, salinity on the PSU scale corresponds closely to parts per thousand.

Evaporation removes fresh water from the surface of the ocean and so increases the salinity of the water left behind. The salinity of the surface of the open ocean varies between

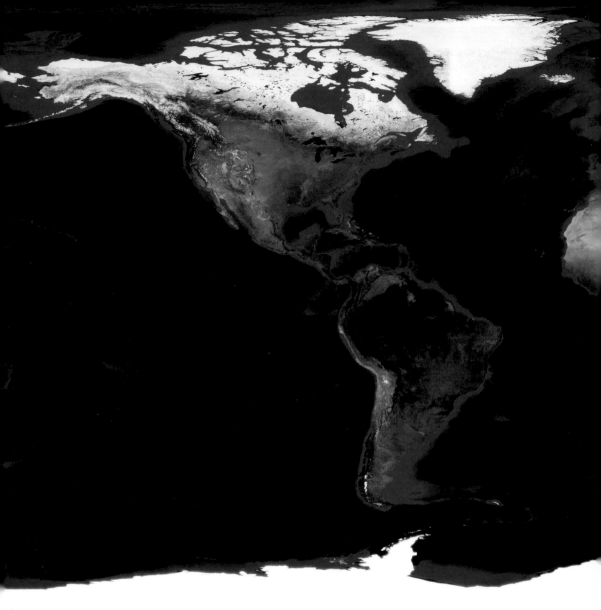

about 34 and 37, with maximum values in the desert latitudes – 30°N and 30°S of the equator, where evaporation exceeds precipitation by the greatest amount (Figure 1.2). The English physicist Robert Boyle was the first to notice the link between ocean salinity and the balance between evaporation and freshwater input. In coastal waters, the salinity variation can be much greater, falling to less than 10 in semi-enclosed seas with high freshwater input

Figure 1.1 Map of ocean depths. The deep blue areas show the deeper parts of the oceans and the lighter blue the shallower parts.

such as the Baltic Sea and rising to over 40 in seas with high evaporation such as the Red Sea.

The vertical variation of salinity can be measured by lowering an instrument called a CTD (which stands for Conductivity–Temperature–Depth) into the ocean (Figure 1.3). Further details of this and other oceanographic instruments are given in chapter 14. The change in salinity with depth at a hydrographic station in the

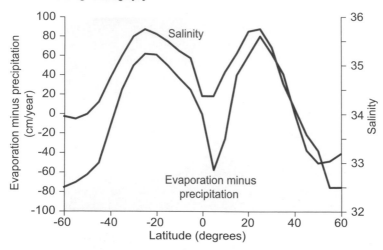

Figure 1.2 Variation of ocean surface salinity (blue line) with latitude. Latitudes south of the equator are shown negative. Maximum salinity occurs in the 'desert latitudes' at approximately 30° North and South. The red line shows the difference between annual evaporation and precipitation. High surface salinity corresponds with high evaporation compared to rainfall.

Figure 1.3 (**A**) The white instrument in the centre of this array is a conductivity–temperature–depth probe (or CTD) being deployed. (**B**) However, normally it is hidden in the middle of a rosette of sampling bottles.

north-east Atlantic is shown in Figure 1.4A. Starting at the surface, the salinity first falls and then rises to a maximum in a layer centred at a depth of 1000 m. This layer of salt water is encountered through much of the north Atlantic and is caused by warm, salty Mediterranean water flowing out through the Strait of Gibraltar. Below this layer the salinity decreases slowly towards the ocean floor. The decrease in salinity with depth is a feature of Atlantic waters and is not observed in the Pacific and Indian Oceans.

The ocean is warmed at the surface by the sun. Surface temperature is therefore greatest at the equator and decreases towards the poles at a gradient of about 1/3 °C per degree of latitude (or about 0.003 °C/km). The sun warms a surface mixed layer a few tens of metres thick, and below this temperature decreases, quickly at first and then more slowly, with depth (Figure 1.4B). The region where temperature is changing is called the main ocean **thermocline**. Here the vertical gradient of temperature can be 10 °C/km: changes in temperature and other properties of ocean water are generally much greater in the vertical than they are in the horizontal. At the bottom of all the oceans, the water is very cold indeed even at the equator. This fact puzzled early oceanographers. In a non-moving ocean, the sun's heat would gradually diffuse down to the ocean floor, producing oceans of nearly uniform temperature from top to bottom. The finding that the bottom waters of the ocean are extremely cold was the first clue in the discovery of the deep, slow movement of ocean waters, now called the **thermohaline circulation**.

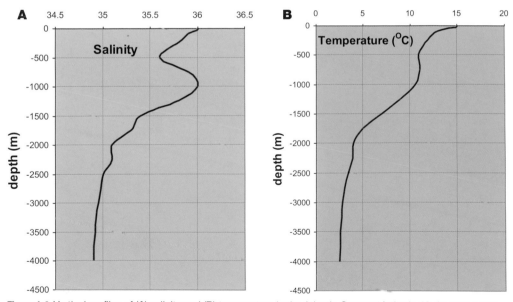

Figure 1.4 Vertical profiles of (**A**) salinity and (**B**) temperature in the Atlantic Ocean at latitude 40 degrees north. Note the maximum in salinity at a depth of 1000 m and the cold water at great depth.

1.2 Water masses and mixing

A water mass is a body of water formed in a particular place in the ocean, and which has characteristic values of temperature and salinity associated with that place. An example of a water mass that we have already met is the Mediterranean water flowing into the Atlantic through the Strait of Gibraltar. When a water mass flows in the ocean it mixes with other water masses, both above and below it as well as at its sides. As a result, it gradually loses its characteristic features. But, before this happens completely, the water masses can be traced for thousands of kilometres, and a picture of the way the deep circulation of the ocean works can be built up. In the case of the Mediterranean outflow, it spreads out through the north Atlantic at a depth of about 1000 m, mixing with Atlantic water both above and below it (Figure 1.5).

Oceanographic observations of temperature and salinity can be plotted on a figure called a **temperature–salinity (or T/S) diagram**, which is a graph of temperature at a given depth against salinity at the same depth. In Figure 1.6, we have plotted the temperature and salinity data from Figure 1.4 in this way. The details of the depth of the measurements are now lost but can be added by annotating some of the points as we have done in this figure. A water mass with a given temperature and salinity can be marked as a single point on a T/S diagram. For example, the characteristic values of Mediterranean Outflow Water shortly after leaving the Strait of Gibraltar are T=11, S=36.5, and we have marked this water mass (labelled as MoW) on Figure 1.6. You can see that the observations in the Atlantic (which were taken 1000 km from the Strait of Gibraltar) tend towards this point at one stage, but don't reach

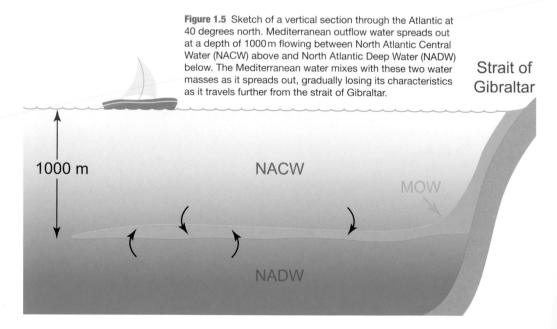

Figure 1.5 Sketch of a vertical section through the Atlantic at 40 degrees north. Mediterranean outflow water spreads out at a depth of 1000 m flowing between North Atlantic Central Water (NACW) above and North Atlantic Deep Water (NADW) below. The Mediterranean water mixes with these two water masses as it spreads out, gradually losing its characteristics as it travels further from the strait of Gibraltar.

Strait of Gibraltar

1000 m

NACW

MOW

NADW

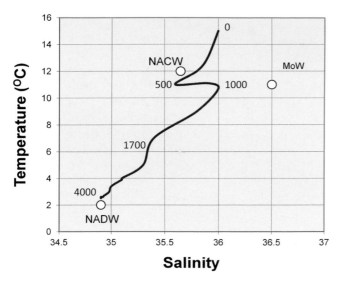

Figure 1.6 A Temperature–Salinity (or T/S) diagram for the data in Figure 1.4. Temperature is plotted against salinity, and the depths of some of the measurements are marked. The characteristic temperature and salinity values of three water masses: Mediterranean Outflow Water, North Atlantic Central Water (in this part of the Atlantic) and North Atlantic Deep Water have been marked on the diagram.

it because of mixing with Atlantic water. We have also marked the temperature and salinity characteristics of the Atlantic Water above and below the Mediterranean outflow on this diagram. These water masses are called North Atlantic Central Water (abbreviated to NACW) and North Atlantic Deep Water (NADW).

A useful feature of T/S diagrams is that, if two water masses mix, the temperature and salinity of the mixture fall on a straight line joining the two water masses on the diagram. This is illustrated in Figure 1.7A, where we have shown the mixing of two water masses, WM1 and WM2. A mixture of 25% WM1 and 75% WM2, for example, would lie ¾ of the way along the line joining WM1 to WM2. The greater the proportion of WM2 in the mixture, the closer this point will get to WM2. If a water mass mixes with two other water masses (as is the case with the Mediterranean outflow) the mixture will lie within a triangle on a T/S diagram (Figure 1.7B). Each of the three water masses occupies a corner of the triangle and

any mixture of these three water masses will have temperature and salinity values lying within the triangle. For example, lines representing different proportions of WM3 are shown in Figure 1.7b. In this way, the proportion of a water mass in a sample of water measured at sea can be calculated if the water masses that have mixed to create it are known.

Figure 1.8 shows the mixing triangle method applied to the data from the north Atlantic station. We have now completed the mixing triangle between the three water masses MoW, NACW and NADW and marked the proportions of MoW within the triangle. It can be seen that the water at a depth of 1000 m at this station contains just over 50% Mediterranean Outflow Water. In travelling 1000 km from the Strait of Gibraltar, the outflow has been diluted by nearly one half as it mixes with Atlantic water above and below. With a little further analysis it is possible to work out the rate at which mixing happens in the ocean in this way.

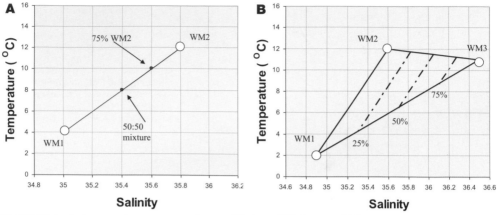

Figure 1.7 Mixing between water masses can be quantified on a T/S diagram. (**A**) Two water masses (WM1 and WM2) are mixing with each other. The temperature and salinity of the mixture lies on a straight line between WM1 and WM2. (**B**) Three water masses are mixing. The temperature and salinity of the mixture lies within a 'mixing triangle'. The dashed lines show different proportions of WM3 in the mixture.

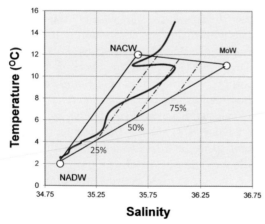

Figure 1.8 Application of the mixing triangle to the data in the north Atlantic. The most saline water, at a depth of 1000 m at this station, contains just over 50% Mediterranean Outflow Water.

1.3 The deep circulation of the ocean

The great value of the water mass concept has been in building up a descriptive picture of the circulation of the oceans. The currents in the deep part of the ocean are extremely slow and too variable for the pattern of the average currents to be determined by direct observation, but water mass analysis allows depths and directions of the currents (but not speed) to be determined. To estimate the speed, we have to add another measurement (for example, dissolved oxygen) to temperature and salinity.

The ocean is warmed by the sun at the surface. Because warm water is buoyant, this produces a warm surface layer in low and temperate latitudes. The ocean is fundamentally different to the atmosphere in this regard. The atmosphere is warmed from below by the heated surface of the earth, and this causes the warm air to rise and create the atmospheric circulation. In the case of the ocean, *warming* by the sun cannot create a circulation in this way: the warm water just sits on the surface. Instead, in the ocean, *cooling* at the surface is the important process. At high latitudes, the ocean gives more heat to the atmosphere than it gains from the sun, and there is a net cooling of surface waters. The cooled water sinks and may be sufficiently dense to replace deeper waters. This cold water, sinking near the poles, spreads out through the bottom of the ocean, creating the deep cold water seen in Figure 1.9. This is the reason why the bottom of the ocean is so cold. Cold bottom waters at the equator have come from the polar seas. It is now known that the sinking takes place in just a few rather localised areas: the Weddell and Ross seas in the Southern Ocean (Antarctica) and near Greenland in the Arctic.

This sinking of water at high latitudes has to be balanced by a rising (or upwelling) of water elsewhere in the ocean. It appears that a slow upwelling of water takes place through-out the rest of the ocean, so that water sinks in just a few localised places, rises back to the surface throughout the ocean and then flows polewards in surface currents, completing the loop. This large-scale circulation of the ocean is called the thermohaline circulation because it is driven by differences in temperature and salinity. Surface heating and cooling alone are not enough to drive this circulation. Energy is also needed, because the kinetic energy created by the sinking is not available to produce the increase in potential energy when the cold waters rise back to the surface (a concept now called Sandstrom's theorem, named after the Swedish oceanographer Johan Sandstrom). It is probable that the energy is provided by friction acting on tidal flows within the main ocean thermocline.

Temperature and salinity sections through the Atlantic from the Antarctic Ice edge to the

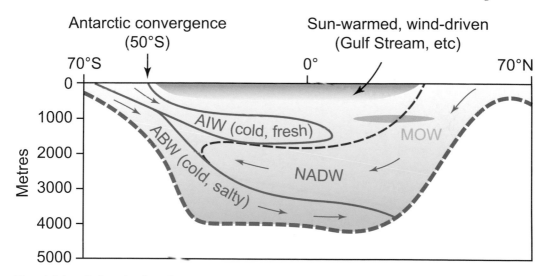

Figure 1.9 A vertical section through the Atlantic from 70°S to 70°N showing the principal water masses identified from their temperature and salinity. The most dense water in this section is Antarctic Bottom Water, formed around Antarctica in winter; this fills the deep basins in both the south and north Atlantic. Antarctic Intermediate Water, formed in summer, is less dense and lies on top of North Atlantic Deep Water, formed in the Arctic. The Mediterranean outflow forms a core at a depth of 1000 m through much of the north Atlantic.

edge of the Arctic Ocean show how the circulation of the deep ocean can be interpreted from T–S data (Figure 1.9). Starting at the southern end of the section, the coldest water (formed in the Weddell Sea in winter) sinks to the ocean bottom and spreads northward as very cold, relatively salty, Antarctic Bottom Water (ABW). Salt is added to seawater during ice formation, and this helps to increase the density of this water. The surface layers of the ocean around the Antarctic continent formed in summer are cold, but less saline because of melting ice. These stretch northward and sink at the Antarctic convergence (about 50 °S) as Antarctic Intermediate Water (AIW). Between these two water masses, North Atlantic Deep Water (NADW) flows southwards. NADW forms in the extreme north Atlantic where cold water sinks to a depth of about 3000 m and travels southwards towards the Antarctic. The distinctive core of Mediterranean outflow water (MoW) is noticeable in both sections of temperature and salinity. This deep circulation is important to ocean life because it carries oxygen down to the very bottom of the ocean, without which the oxygen in the depths would be depleted, or even used up, by the decomposition of organic matter (*see* chapters 8 to 11).

There is no equivalent to the sinking of North Atlantic Deep Water in the Pacific Ocean. Surface waters in the north Pacific are somewhat cooler than those at the same latitude in the Atlantic. This reduces the evaporation rate, and north Pacific waters have a lower salinity than those in the Atlantic. As we shall see in the next chapter, the density of seawater is particularly sensitive to salinity. The lower salinity of the north Pacific prevents the water from becoming sufficiently dense to sink to the ocean floor.

2   Density and density flows

The **density** of an object or a parcel of fluid is equal to its mass divided by its volume. The density of seawater is an important concept in oceanography because dense water sinks below less dense water, creating currents and layering in the ocean.

The density of fresh water depends on temperature, and at 15 °C is approximately 1000 kg m⁻³. Seawater is a little denser than fresh water at the same temperature, as we shall see. The human body is made mostly of water, so we would expect the density of our bodies to be about the same. In fact, the average density of the human body is about 950 kg m⁻³, and so most humans float on water (are buoyant) because they are less dense than seawater.

2.1 Archimedes' Principle

Archimedes' Principle deals with the **buoyancy** objects experience when they are immersed in water. Things weigh less in water than they do in air. This is because they experience an upthrust equal to the weight of water they have displaced (they are supported by the surrounding water in just the same way that the water they have replaced was). This is a statement of **Archimedes' Principle**. If a body has a mass m, it has a weight in air equal to mg, where g is the acceleration of gravity (equal to 9.81 m s⁻²). The same object in water will weigh mg where g is called the **reduced gravity** (Figure 2.1). We can calculate the reduced gravity g with the formula:

$$g' = g\frac{d_2 - d_1}{d_2}$$

where d_2 is the density of the submerged object and d_1 is the density of water. For example, a grain of sand which is three times as dense as water will feel a reduced gravity ( )g when it is suspended in seawater. Its weight in water will therefore be of its weight in air.

A special case of Archimedes' principle applies to floating bodies, when the principle states: '*a floating body displaces its own weight of water*'. By 'displaces' we mean 'pushes out of the way'. So if you float an object on the surface of water, it pushes out of the way a volume of water that has the same weight as the floating object.

An iceberg weighing, say, 1 tonne (1000 kg), will have a volume of 1087 m³ (the density of ice is about 920 kg m⁻³, and since density =

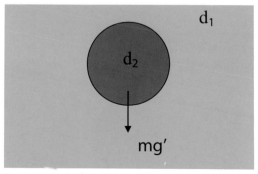

Figure 2.1 Illustration of reduced gravity. An object of mass m and density d_2, surrounded by water of density d_1 has a weight mg' where g' (defined by equation 2.1 in the text) is called the reduced gravity.

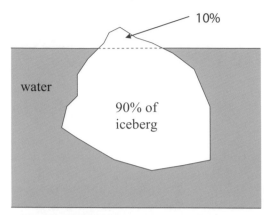

Figure 2.2 A floating iceberg displaces its own weight of water. The volume of the displaced water is about 90% of the volume of the iceberg; the remaining 1/10th of the volume of the iceberg is visible above the ocean surface. The photograph shows the tip of a floating iceberg.

mass/volume it follows that volume = mass/density). If you place this iceberg in the sea, it will float because it is less dense than water, and it will displace 1000 kg of water exactly (its own weight). 1000 kg of seawater occupies about 976 m³, so some of the iceberg will stick out above the water surface. 976 m³ (about 90%) of the iceberg will be under water, and the remaining 111 m³ (about 10%) will stick above the water. When you see floating ice (not pack ice or frozen seawater) at sea, you only see 'the tip of the iceberg' (Figure 2.2).

What would happen if all the icebergs at sea were to melt? If we consider the 1 tonne iceberg, it has pushed 1000 kg of water out of the way, so it will cause sea level to rise by the equivalent of 1000 kg of water. When the iceberg melts, it will no longer displace any water but it will add 1000 kg of water to the sea (the mass will stay the same on melting). This is illustrated with the three photographs of a simple laboratory experiment shown in Figure 2.3. The first photograph (a) shows a tank containing some fresh water. The second photograph (b) shows the same tank with some small pieces of ice added. The water level rises as the floating ice displaces its own weight of water. Note that only a small part of the ice lies above the water surface. In the

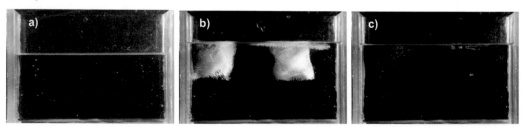

Figure 2.3 Ice added to water in a flask raises the water level as water is displaced to support the floating ice (**a** and **b**). When the ice melts (**c**) there is little change in water level because the mass of melted ice water is the same as the mass of displaced water.

third photograph (c), the ice has melted and the water level is the same as in (b) because the mass of water added by the melting ice is the same as that which has been displaced by the floating ice.

A consequence is that the melting of floating icebergs produced by global warming will not have a big impact on sea level. The melting of land-based glaciers is a different matter altogether. Land-based glaciers such as those on Greenland and the Antarctic continent are not currently displacing any seawater, so when they melt they will add a new mass of water to the ocean. It is estimated that the rise in sea level produced by the melting of all the land-based ice will be 80 m.

2.2 The density of water

Water is heavy. Its density is approximately $1000\,kg\,m^{-3}$ and varies with temperature, salinity and (to a small extent) pressure. Fresh water has a maximum density at 4 °C and this fact is important in preventing ponds and lakes from freezing solid in winter. When surface water cools to 4 °C it sinks to the bottom of the lake. Further surface cooling produces water that is less dense, and this water therefore floats on top of the 4 °C water at the bottom of the lake. The surface water may eventually cool to the point where ice forms, but the fact that there is liquid water at the bottom means that animals and plants can survive the winter in freshwater lakes.

No such phenomenon occurs in the ocean. The density of seawater continues to increase as the temperature decreases. Salinity is also an important control on seawater density, which increases with increasing salinity. Mostly because of the salt content, the density of seawater is somewhat greater than that of fresh water. Typically, the density of seawater is a little greater than $1000\,kg\,m^{-3}$. Oceanographers call the difference between the density of a sample of seawater and $1000\,kg\,m^{-3}$ 'sigma-t', given the symbol σ_t. So, for example, seawater that has a density 1022.5 $kg\,m^{-3}$ has a σ_t value of 22.5.

2.3 'sigma-t' on temperature salinity diagrams

Because the density of seawater depends mostly on temperature and salinity, lines of constant density can be marked on T/S diagrams. Figure 2.4 show how these look. For example, water which has a temperature of 6 °C and a salinity of 35 has a σ_t of 27.5 and therefore a density of $1027.5\,kg\,m^{-3}$. Note that the σ_t contours become quite curved at low

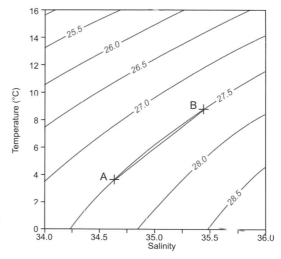

Figure 2.4 Temperature/Salinity diagram with lines of σ_t added. Note the curvature of the constant density lines at low temperature. If two water masses A and B with the same density are mixed, the mixture will have a greater density. This effect, called caballing, is thought to be important in the formation of the densest bottom waters of the oceans.

temperature. This can lead to an effect called **caballing**, which is thought to be important in the production of dense water at high latitude. In Figure 2.4, the two water masses A and B have the same density. If they mix, the mixture will lie on the line joining them. The density of the mixture will therefore be greater than either A or B and therefore the mixture will sink below the two water masses out of which it has formed.

2.4 Layering of water

All water, including seawater, likes to find its own level with the densest water on the bottom. This layering is called **stratification** and is a physical property of seawater that has important biological and chemical consequences.

We have already come across one important form of layering in the ocean, the main thermocline, which can be several kilometres thick and which separates the sun-warmed surface layers from the deep cold waters. Another form of stratification is found in estuaries and coastal waters, where a layer of relatively low salinity water, produced by rivers flowing into the sea, lies on top of a layer of salty water. The interface between the low and high salinity water is called a **halocline**. Stratification can therefore be produced by salinity and temperature, and sometimes by both acting together. For stable stratification to exist, the densest water will always be on the bottom. The interface between waters of different density is called a **pycnocline**.

When water is strongly stratified, it is possible for waves, called **internal waves**, to form on the interface between the different layers. We cannot normally see these waves from above the surface, but their effects can be

measured with instruments recording temperature and salinity at a given depth. The vertical movement associated with the internal waves can be tens of metres – much bigger (and also slower) than surface waves. Submarines that have set their buoyancy to cruise at a particular depth (and therefore density level) are carried up and down with this movement. In highly stratified regions such as fjords, surface motor vessels can experience a phenomenon called 'dead water'. The energy from their propeller goes into creating internal waves instead of driving the boat forward (Figure 2.5).

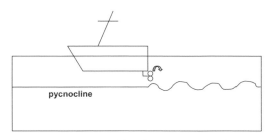

Figure 2.5 Illustration of the 'dead water' effect. The propeller of a motor boat in a sea with light surface water separated from denser deeper water by a pycnocline makes internal waves on the pycnocline. The energy from the motor goes into creating these waves instead of propelling the boat forward.

2.5 Density currents

When two water masses of different density lie adjacent to each other, there is a tendency for the denser of the two to flow under the less dense (and vice versa) to create a circulation known as a **density current**. Density currents (also called gravity currents) occur in may forms in the ocean. For example, the thermohaline circulation discussed in Chapter 1 is essentially a large-scale density current. In estuaries, salty ocean water can flow under the less dense river water. Other forms of density current include **turbidity currents** in which particles suspended in water increase the density. In the atmosphere, avalanches are also a form of density current in which suspended snow has increased the density of the air.

Density currents can be created and studied in a laboratory tank experiment known as the **lock gate experiment** (Figure 2.6). A perspex tank is set up with salt water at one end, divided by a removable lock gate from fresh water at the other. The water can be coloured with a dye to make the subsequent flow easy to see. When the lock gate is withdrawn, the salty water flows under the fresh water along the bottom of the tank and the fresh water flows over the salt water in the opposite direction. With salinities typical of seawater, the flow is quite slow – typically just a few cm s^{-1} for a tank 20 cm deep. Physically, the sinking of the salt water below the fresh produces a reduction in potential energy, and the kinetic energy of the resulting flow is derived from this. By matching the loss in potential energy to the

Figure 2.6 The lock gate experiment. Salt water (coloured blue) is separated from fresh water by a lock gate. When the lock gate is removed, the salt water flows underneath the fresh in a current called a density current. Friction at the interface between the flowing salt water and the overlying fresh water causes internal waves to form on the interface.

gain in kinetic energy, it can be shown that the speed of the resulting density current depends on just two things: the density difference between the salt and fresh water and the depth of water in the tank. We can write an equation for the speed, S, of the current as follows:

$$S = C\sqrt{g'h}$$

Here, g' is the reduced gravity (*see* section 2.1) and h is the thickness of the current. The factor C depends on the geometry of the tank: it is about 1 but in general must be determined exactly by experiment. In section 2.1 we defined reduced gravity in terms of an object immersed in water, but it can also be applied to two liquids of different density. For example, for the experiment shown in Figure 2.6, the density of the salt water was $1020\,kg\,m^{-3}$ and that of the fresh water was $1000\,kg\,m^{-3}$, giving $g'=9.81$ x $(1020–1000)/1020=0.19\,m\,s^{-2}$. The depth of the current in Figure 2.6 is 5 cm (0.05 m) and its speed was measured as $0.07\,m\,s^{-1}$. Substituting these values into the equation above gives $C=0.5$.

2.6 Density and pressure

The **pressure** at a surface in the ocean is equal to the weight of water above that surface (expressed as a force per unit area). The weight of water will equal the average density of the water times its volume times the acceleration of gravity. Because seawater is heavy compared to air (it is about 1000 times denser than air) the pressure in the ocean is much greater than the pressure in the atmosphere. As a rule of thumb, the pressure in the ocean increases by 1 atmosphere for every 10 m increase in depth (1 atmosphere is equal to the atmospheric pressure at sea level, also called 1 bar). The pressure in the ocean at a depth of 1000 m,

therefore, will be 101 atmospheres compared to 1 atmosphere at the sea surface. An increase in depth by 1 m creates an increase in pressure of 1/10th of a bar, or 1 **decibar**.

It is a feature of pressure in a fluid that the pressure at a point acts equally in all directions. If we put a pipe down into the ocean, say to a depth of 1000 m, the pressure at the bottom of the pipe will be 101 bars compared to 1 bar at the top of the pipe. This difference in pressure just balances the weight of the water in the pipe. Now imagine that we displace the water in the pipe upwards (perhaps by applying suction at the top of the pipe). Imagine we are at a place where temperature decreases with depth (which is normal) and where salinity also decreases with depth (as it does in the Atlantic). In that case, we will suck cold, fresh (and dense) water into the pipe, compared to the water outside the pipe (*see* Figure 2.7). Before the water can sink back down again, it is possible that heat will flow into the pipe and the temperature will equilibrate with that outside. But the water in the pipe will still be fresh compared to that outside (salt can't flow through the pipe). Because fresher water is less dense, the water in the pipe will weigh less than that outside, and it will no longer balance the pressure difference between the ends of the pipe. As a result it will be forced upwards, the cycle will repeat, and a continuous flow will be produced. This 'salt fountain' (first proposed by the American oceanographer Henry Stommel) is a potential way of generating energy from the ocean.

A smaller scale version of the salt fountain can be created in a laboratory tank. A tank is set up with warm salty water overlying cold fresher water. The difference in temperature is sufficient to make the lower layer denser, and

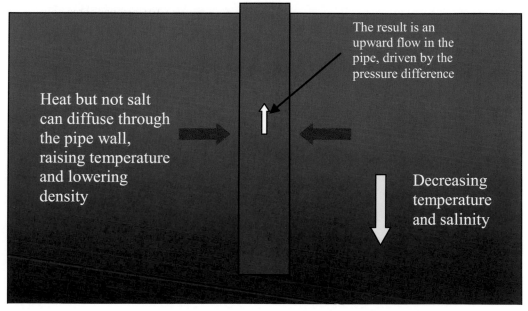

Figure 2.7 The salt fountain. An initial suck on the pipe creates a continuous upwards flow of water, which derives its energy from the vertical temperature and salinity gradients in the ocean.

so the layering is initially stable. At the interface between the two layers, the salt and heat diffuse by molecular processes. The molecular diffusion of heat is faster than it is for salt, and so the upper layer cools more quickly than it freshens. Similarly the lower layer warms more quickly than it becomes saltier. This creates local instabilities, with denser water lying on top of less dense water. As a result, water movements occur in small columns (called **salt fingers**) extending both upwards and downwards. The salt fingers penetrate only so far before lateral diffusion of salt destroys the density difference. It is thought that this process may create layers in the ocean with sharply defined interfaces of temperature and salinity. Sometimes **temperature staircases** (in which the temperature varies with depth in a series of steps) are observed in the ocean and are thought to be produced by this mechanism.

3   Ocean waves

Ocean waves are a familiar sight to anyone who has looked at the sea. Waves are important because they may be a hazard to shipping, and they carry energy that has the potential to be harnessed to our benefit. Waves near the beach make swimming in the sea fun, and the whole surfing industry is based on good beach waves.

On a calm day, waves can approach the beach as a regular sequence of parallel humps of water. These waves have travelled from a distant storm, often over many kilometres of

Figure 3.1 Swell waves arriving at a beach. Waves like this have travelled from a distant storm. Because longer waves travel faster, it is the longest waves from the storm that arrive at the beach first.

Figure 3.2 Waves made in a coastal sea by the action of the wind are short and choppy, often travelling in more than one direction.

ocean, and are called **swell** waves (Figure 3.1). On more windy days, swell waves can be supplemented by waves generated locally, and which have a more confused pattern, sometimes called '**sea**' (Figure 3.2).

3.1 Making waves

Waves can be made in a pond by lobbing in a pebble. The nearest oceanic equivalent to this is a **tsunami**, which is a wave created by an underwater earthquake. In both the pond and the tsunami, the waves travel outwards from the disturbance into the surrounding water.

Waves can also be created in a channel by moving a paddle back and forth (Figure 3.3). This creates a regular train of waves travelling in one direction that is useful for measuring the effect of waves on, say, sand transport. The most familiar form of water waves, though, is as **wind waves**, created by the action of the wind on the water surface. It is not easy to understand how an essentially horizontal wind can create vertical waves, but the friction between

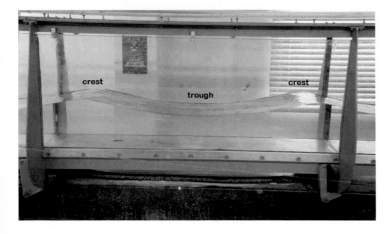

Figure 3.3 (top) Waves can be created in a long tank (a flume) by oscillating a paddle at one end. The wavelength is the horizontal distance between two crests (bottom picture) and the wave height the vertical distance between a crest and a trough.

the air and the water creates a rubbing force in the same way that you can create crinkles in a tablecloth by moving your hand over it.

Any water wave has a number of features. The highest point in the wave is called the wave crest and the lowest the wave trough. The vertical distance between the crest and the trough is the **wave height**. The horizontal distance between two crests is called the **wavelength** (L). The ratio of wave height to wavelength (called the wave steepness) seems to have a maximum value of about 1/7; waves steeper than this break. If you watch the crest of a wave, it travels along at a speed that is called the **phase speed** (usually denoted by the symbol c). The time between one crest and the next passing a fixed point is called the **wave period** (T). Since the wave has travelled exactly one wavelength in this time, it follows that $c = L/T$.

Below the waves, the water moves in wave-induced currents called **orbitals**. The speed of the water in these currents is different to the speed of the waves. We will return to the motion of the water in the orbitals in section 3.3.

3.2 Energy in waves

It takes energy to make waves, and the bigger the waves are, the more energy is needed. In fact it can be shown that the energy in waves is proportional to the square of the wave height (Table 3.1). Even moderate waves have a lot of energy, and the energy travels with the waves. There are many ingenious devices for trapping this energy and turning it into useful electricity and pilot and commercial wave energy plants are becoming established. A commercial wave energy plant is in operation on Islay in Scotland.

When the wind blows over the ocean it exerts a frictional force on the water surface.

Table 3.1: The energy in waves is proportional to the square of the wave height.

Wave height (m)	Energy per unit area Joules m^{-2}
1	1,226
2	4,905
5	30,656
10	122,625
20	490,500

There is a rule in physics that says that the work, or energy, done by a force is equal to the force times the distance it moves. If you have a steady wind blowing over the sea, the energy it puts into the sea will therefore be proportional to the distance over which the wind acts upon the water. This distance is called the **fetch**. The energy in wind waves should therefore be proportional to the fetch. Since the energy is also proportional to the square of the wave height, we might expect that the square of the wave height is proportional to the fetch, or (alternatively) that the wave height is proportional to the *square root* of the fetch. This indeed seems to be the case (as was realised by Thomas Stevenson, father of Robert Louis). Experimental evidence suggests that:

$$H = \frac{2}{3}\sqrt{F}$$

where H is the wave height in metres, and F is the fetch in kilometres. Using this formula, we can see that in a sea 100 km across, we would expect the largest locally generated waves to have a height of about 8 m. In the Atlantic Ocean, with a fetch of 6000 km, we might get waves up to 50 m high. Waves in the Pacific Ocean, which is wider than the Atlantic and therefore has a greater fetch, could be bigger than this. The largest fetch of all is in the

Southern Ocean surrounding Antarctica, which has no end, and therefore a potentially limitless fetch. The waves in the Southern Ocean are notorious and the Antarctic explorer Sir Ernest Shackleton, who made a famous small boat journey across this ocean, gave a vivid account of huge waves in his book *South*:

> 'So small was our boat and so great were the seas that often our sail flapped idly in the calm between the crests of two waves. Then we would climb the next slope and catch the full fury of the gale where the wool-like whiteness of the breaking water surged around us.'

The fetch is an important influence on wave height, but it is not the only one. The wave height will also depend on the strength of the wind, and the time for which it has been blowing (a parameter called the duration in wave forecasting). Modern practical wave forecasting techniques therefore take account of three things: fetch, wind strength and duration. You can find a copy of wave forecasting diagrams in, for example, *Oceanography and Seamanship* by William G. Van Dorn.

3.3 Wave propogation

Once waves are made, they travel (or propagate) away from the generation area. The way they do this is that the crests and troughs travel along, and the water oscillates back and forth as the wave passes through. It is the *shape* of the wave that travels. The *speed* of travel and the way that the water oscillates, depends on how deep the water is compared to the wavelength, so we shall treat the two cases separately.

Deep-water waves

When the water is deep compared to the wavelength, it is observed that water particles go round in almost closed circles (called orbitals) as the waves travel past (Figure 3.4). These circles are largest at the surface, where they have a diameter equal to the wave height. The

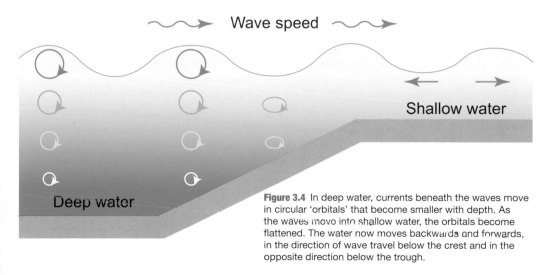

Figure 3.4 In deep water, currents beneath the waves move in circular 'orbitals' that become smaller with depth. As the waves move into shallow water, the orbitals become flattened. The water now moves backwards and forwards, in the direction of wave travel below the crest and in the opposite direction below the trough.

orbitals get smaller as you go deeper, until one wavelength down they have virtually disappeared. For this reason, submarines can dive to avoid storm waves. The maximum speed of the water movement in the orbitals is equal to the diameter of the largest circle (ϖH) divided by the wave period. For 1 metre high waves with a period 10 seconds, this is 31 cm s^{-1}. The speed at which the waves travel is much faster. For deep-water waves, it can be shown that the wave speed is proportional to the period (Table 3.2).

Table 3.2: For deep-water waves, it can be shown that the wave speed is proportional to the period.

Wave period (seconds)	Wave speed in deep water or phase speed (m s^{-1})
1	1.56
2	3.12
5	7.80
10	15.6
15	23.4

Because waves with long period go faster, and the wavelength is equal to the wave speed times the period, it follows that waves with long wavelength travel faster. When waves are made in a storm, they travel outwards across the ocean, and the longer waves run ahead and separate out. This effect is called **wave dispersion.** This is the reason why swell waves, arriving at a shore from a distant storm, have such a regular appearance. In travelling across the ocean, waves with different wavelengths separate out. The longest waves arrive at the shore first and gradually, over a period of days, shorter wavelength waves will appear. We can use this effect to work out how far swell waves arriving at a beach have travelled (*see* text box). You can observe a similar effect on the banks

Working out how far swell waves have travelled

One day, swell waves with period 15 seconds are observed to arrive on a beach. The next day at the same time, the period of the swell waves has shortened to 10 seconds. How far away is the storm that generated this swell?

The speed of 15 second waves is 23.4 m s^{-1} (84 km h^{-1}); the speed of 10 second waves is 15.6 m s^{-1} (56 km h^{-1}). Let the distance to the storm be x km. Since time = distance/speed, and the difference in time between these waves arriving is 24 hours, we have

$$\frac{x}{56} - \frac{x}{84} = 24$$

This can be solved for x (by multiplying both sides by 56 times 84) to show that x is 4032 km. So the waves have travelled a long way. In fact, they've probably travelled even further than that because of an effect called the group velocity (see main text).

of a river when a speedboat goes past, and the wake arrives on the shore. In this case, the waves hitting the shore will be audible lapping on the shore, and it is noticeable that the time between 'laps' gets shorter as time goes on.

The speed of the waves shown in Table 3.2 is the speed of a wave with exactly the given period – it is called the phase speed. However, swell waves at sea actually consist of a number of waves of similar period travelling together, called a wave group. You can make a wave group in a tank such as that shown in Figure 3.3 by switching the generating paddle on for a few seconds and then switching it off; you will create a group of waves which travel down the tank away from the paddle. If you watch the group carefully you will see that the wave leading the group disappears, and a new one grows at the back. This only happens in deep-water waves. What happens is that the wave energy cannot travel as fast as the phase velocity of the waves – in fact it can only travel half as fast. So the leading waves run out of energy, and this energy is passed to the waves at the back. The speed of travel of the wave group is called the **group velocity**, and for deep-water waves this is just one half the deep-water wave phase speed. Because of this, the distance to the storm is twice as far as we estimated in the text box – about 8000 km. If they were measured in Britain, they would have come from as far as the south Atlantic.

Shallow-water waves

When waves move into water that is shallow compared to their wavelength, the orbitals become squashed. Now, as the wave goes past, the water particles oscillate backwards and forwards almost horizontally (Figure 3.4). Under the crest of the wave, the current flows in the same direction as the wave, and under the trough it travels in the opposite direction. The strength of the current increases with wave height and increases as the water gets shallower. For these shallow-water waves both the phase speed (that is the speed at which the crest advances) and the group velocity (the speed at which the wave energy advances) are the same and depend only on the water depth. The speed can be calculated from the formula

$$c = \sqrt{gD}$$

where c is the wave speed (in metres per second), g is the acceleration of gravity and D is the water depth (in metres). Some wave speeds in different water depths are shown in the table. You may wonder why a depth of 4000 m is included for shallow-water waves. Remember, we are talking about water that is shallow *compared to the wavelength,* and for very long waves (tides and tsunamis) even the deep ocean can appear shallow. Notice that these waves can travel *very fast.* We have included in this table the maximum current speed under the wave assuming the wave is 1 m high. (For other wave heights, multiply these figures for the maximum current speed by the wave height, so for 2 m waves, you would double these figures). Notice that the current speed is usually much less than the wave speed, but in fact the two become equal when the water depth is about the same as the wave height. At this point the currents in the wave are going as fast as the wave is travelling, and the wave tends to break (Figure 3.5).

Table 3.3: Relationship between depth, wave speed and maximum current for shallow-water waves.

Water depth (m)	Wave speed (m s⁻¹)	Wave speed (km h⁻¹)	Maximum current speed for waves 1 m high (m s⁻¹)
1	3.1	11.3	3.1
10	9.9	35.7	0.99
100	31.3	112.7	0.31
4000	198.1	713	0.05

3.4 Waves near the shore

The fact that, in shallow water, the wave speed just depends on the water depth leads to a number of interesting phenomena, one of which is **refraction**. If waves approach a shore at an angle, the part of the wave crest closest the shore travels more slowly than that further offshore, and as a result the wave crest becomes curved. The angle that the crest makes with the shore obeys a law from optics called **Snell's Law**. This says that the sine of the

Figure 3.5 Waves approaching the shore tend to break when the wave height is about the same as the water depth. In this photograph, the height of the wave at the back is seen to increase from left to right (because the water is getting shallower in this direction) and the wave breaks at a point where the ratio of the wave height to the water depth reaches a critical value.

Figure 3.6 Surfing in breaking waves on the coast of Brittany, France. Surfers ride on the face of the breaking wave, where the lift provided by the current flowing towards the crest is sufficient to support their body weight.

angle the crest makes with the shore divided by the speed of the wave is a constant. As the waves move into shallow water and their speed reduces, so does the sine of the angle between the crest and the shore.

Another effect of the waves slowing down as they approach the shore is that they get bigger. Conservation of energy requires the product of the wave speed and the energy in the wave to be a constant. As the speed gets less, the energy must increase, and this means that the wave height increases. This continues until the wave breaks and gives up its energy.

Surfers like to ride waves that are just on the point of breaking and remain so for a long time. A flat beach is best from this point of view, and this sort of beach tends to produce spilling waves. There is a position on the face of the wave that is stable in the sense that the weight of the surfer down the wave face is just balanced by the current moving towards the crest (Figure 3.6). It is a feature of breaking waves that the current is close to the wave speed itself.

4 Flow in the oceans

One of the most fascinating aspects of the ocean is that it always appears to be moving. At the large scale, we experience this movement in the major ocean currents such as the Gulf Stream. Much of the energy in the moving ocean, however, is present on a smaller scale: circulating eddies a few tens of kilometres in size. The energy that drives this movement is ultimately derived from the heat of the sun and the rotation of the earth.

4.1 Making ocean water flow

The great surface ocean currents, for example the Gulf Stream in the north Atlantic and the Kuroshio in the north Pacific, are primarily driven by the wind. Ocean surface winds in the Atlantic and Pacific (both north and south) are, in a general sense, towards the west near the Equator (the trade winds) and towards the east at higher latitudes (the westerlies;

see Figure 4.1). This produces a clockwise circulation in the north Atlantic and the north Pacific, and anti-clockwise circulation in the southern parts of both oceans. These great circular flows in the major ocean basins are called *wind-driven gyres*, a term coined by the American physical oceanographer, Walter Munk. The Indian Ocean is somewhat different and has seasonally varying currents that respond to changes in wind direction associated with the monsoon. It was in fact the response of the Indian Ocean to the wind that helped to convince oceanographers that surface currents were primarily wind-driven. The wind-driven circulation in the Atlantic and Pacific, however, is not symmetrical. The currents on the western side of these oceans are much faster and more intense than those on the east. In the north Atlantic, for example, the Gulf Stream in the west is much more

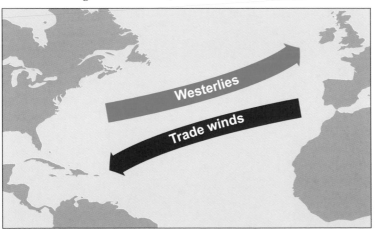

Figure 4.1 The prevailing wind patterns over the north Atlantic drive a predominantly clockwise flow in the surface layer.

obvious than the gentle southward flow that occurs down the coast of Europe and North Africa. The reason why this should be so evaded oceanographers for many years until Henry Stommel, an American oceanographer, proposed a solution in 1948. The **western intensification** of ocean currents is caused by the change in angular momentum of the water as it flows pole-ward (*see* text box).

Figure 4.2 shows a schematic of the major currents of the north Atlantic. Surface currents are marked in red and these show the progress of the Gulf Stream across the north Atlantic. The flow divides at the latitude of Spain, and it is the rather weaker North Atlantic Drift that brings relatively warm water from the south and west to the British Isles. The rest of the Gulf Stream water completes a circuit of the ocean in a great clockwise *gyre*.

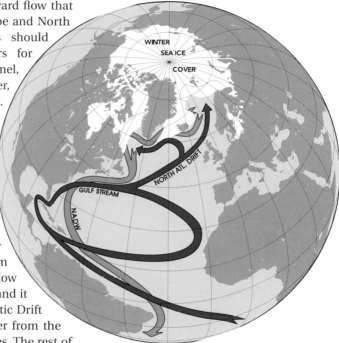

Figure 4.2 Schematic of the circulation in the north Atlantic showing surface currents in red and deep flows in blue.

Why there is a Gulf Stream

In the Pacific and Atlantic Oceans (both north and south) the current on the western side of the ocean is much faster and more intense than anywhere else. We shall explain as simply as possible why this is, using the Gulf Stream as an example.

A vertical column on the surface of the earth possesses angular momentum imparted by the Earth's spin (oceanographers prefer the word *vorticity*, but we shall stick with the hopefully more familiar term angular momentum). The angular momentum is greatest at the poles, where the column makes a complete turn once per day, and is

zero at the Equator. As the Gulf Stream flows northwards up the east coast of America, therefore, it has to acquire anti-clockwise angular momentum to keep up with the increasing effect of the Earth's spin. The winds don't help, because the prevailing winds impart clockwise momentum, and it is this that drives the gyre. Instead, the Gulf Stream acquires anti-clockwise angular momentum through friction. By pressing against the east coast of America, anti-clockwise spinning eddies are generated which provide the necessary angular momentum. In order to provide enough angular momentum to allow for the change of latitude *and* counteract the clockwise momentum put in by the wind, a fast narrow flow is required to generate a lot of friction.

On the eastern side of the ocean, the equator-ward flow needs to lose anti-clockwise momentum. The effect of the clockwise wind will help to do this, and so here the friction is not needed (in fact it will produce anti-clockwise momentum, which is the opposite of what is needed). As a result, the current down the east side of the ocean is broad and slow to minimise frictional effects. To summarise, to preserve angular momentum, poleward flowing currents need a lot of friction with the continental boundary and Equatorward-flowing currents do not.

The second great mover of ocean water is a difference in density. Dense water formed by cooling at high latitudes sinks and contributes to the thermohaline circulation, part of which is shown in blue in Figure 4.2 contributing to North Atlantic Deep Water flowing southwards along the western side of the Atlantic. As we discussed in Chapter 1, a source of energy is needed to drive this circulation, and this is thought to be the tidal energy generated within the body of the ocean. Tidal forces are, in fact, the third of the great forces that can drive flow in the ocean, and we deal with these in Chapter 5.

4.2 Observing ocean currents

Oceanographers observe ocean currents directly in two ways. The first is to use a current meter at a fixed point. In its simplest form a current meter has a propeller that turns in the current; the number of turns of the propeller in a given time is counted and this is proportional to the current speed. The current meter would normally be fixed on a mooring attached to the seabed. Current measurements made at a fixed point in this way are called **Eulerian** observations. Typical current speeds range from about $200\,cm\,s^{-1}$ in strong currents like the Gulf Stream to less than $1\,cm\,s^{-1}$ in the deep parts of the ocean. In shallow water, Eulerian measurements of current velocity can also be made with **Acoustic Doppler techniques** (*see* Chapter 15).

The second way to measure ocean currents is with a drifting buoy that can be tracked by satellite. Drifting buoys follow the current: they don't stay in one place. Measurements of ocean currents made in this way are called **Lagrangian**. A very successful international collaboration on ocean drifters is the Argo collection of a large number (several thousand) of small drifting probes. The probes drift at depths of up to 2 km, returning to the surface to transmit data on water temperature and salinity. This data is made freely available via the internet (Figure 4.3).

The strength of ocean currents can usefully be expressed as the volume of water transported in a given time. The **volume transport** in a current is equal to the flow speed multiplied by the cross-sectional area of the current. Millions of cubic metres per second (m^3/s) are transported by major ocean currents, and we have a special unit for recording this, the **Sverdrup** (symbol is Sv), named after the Norwegian oceanographer, Harald Sverdrup. One Sv is one million $m^3\,s^{-1}$. The volume transports of some selected ocean currents are shown in Table 4.1.

The effect of surface ocean currents can be seen in the distribution of surface properties such as temperature. Currents moving from

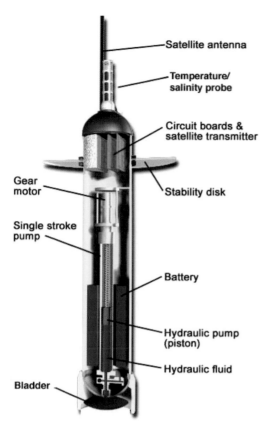

Figure 4.3 Cross section of an Argo float, a free-floating oceanographic probe.

- Satellite antenna
- Temperature/salinity probe
- Circuit boards & satellite transmitter
- Gear motor
- Stability disk
- Single stroke pump
- Battery
- Hydraulic pump (piston)
- Hydraulic fluid
- Bladder

Table 4.1: Volume transports of some selected ocean currents.

	Volume transport (Sv)
The Gulf Stream (off Cape Hatteras)	85
wThe Kuroshio (Japan)	40
Antarctic Circumpolar Current	110

low to high latitudes carry warm water. The warm waters of the Gulf Stream can be made out in infra-red satellite images (Figure 4.4).

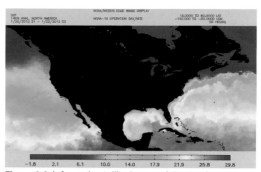

Figure 4.4 Infra-red satellite image of the north-west Atlantic showing sea surface temperature. The path of the Gulf Stream as it leaves the coast of the United States and travels eastwards and northwards towards Europe can be seen.

4.3 The continuity principle

It is not always necessary to measure ocean currents directly. Sometimes the volume transport, the most important aspect of a current, can be inferred by applying a rule called the continuity principle. This rule states, in effect, that 'what comes in must go out'. We will illustrate this principle by applying it to the flows in and out of the Mediterranean Sea through the Strait of Gibraltar.

The Mediterranean straddles latitude 30° N and in general in this area, evaporation exceeds precipitation and run-off from rivers. There is thus a net loss of fresh water from the Mediterranean Sea. The evaporation rate is about 1 metre each year, that is the surface of a pool of water that is not topped up with a hosepipe will fall by $1\,\text{m year}^{-1}$. The surface area of the Mediterranean is about 2.5 million km². The volume of water lost by evaporation in one year is the product of these two figures, namely $2.5 \times 10^{12}\,\text{m}^3\,\text{year}^{-1}$, or $100{,}000\,\text{m}^3\,\text{s}^{-1}$, or about 0.1 Sv. Since this water is lost, it must be replaced (otherwise the Mediterranean would eventually dry out, which it shows no signs of doing). The major connection between the Mediterranean and the rest of the world ocean is through the Strait of Gibraltar. The water lost by evaporation can therefore be replaced by a flow, equal to 0.1 Sv, through the Strait of Gibraltar.

The Strait of Gibraltar is 10 km wide and about 250 m deep at its narrowest and shallowest section, which gives a cross-sectional area of 2,500,000 m². A flow of 0.1 Sv through a channel of this area will have a speed 100,000 /2,500,000 = 0.04 m s⁻¹ = 4 cm s⁻¹. So the flow we might expect is quite slow.

However, it's not as simple as this, because the flow in through the strait comes from the Atlantic and is salty, whereas the water that is lost by evaporation is fresh.

If you replace fresh water by salt water, the Mediterranean will get saltier with time. Again, there is no evidence that this is happening, so there must be some way in which the Mediterranean Sea is getting rid of salt. It does this with a deep flow of extra salty water through the Strait of Gibraltar. Observations in the narrowest section of the strait (Figure 4.5) show that the surface water has a salinity of about 37 and is flowing in. Below this, there is a layer of saltier water, with a salinity of about 38, which is flowing out. We can apply the principle of continuity to the salt as well as the water. This gives us two simultaneous equations for the inflow and outflow. When these are solved (*see* text box and Figure 4.6 for details), we learn that, in order to preserve the salinity *and* the volume of water in the Mediterranean, the inflow must be 3.8 Sv and the outflow 3.7 Sv.

Notice that, because of the requirement to maintain a salt balance, the flow rates are now much greater – in fact the volume flow in is

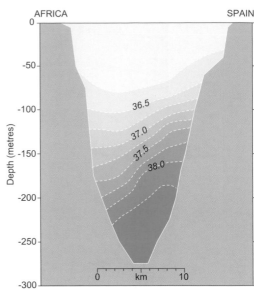

Figure 4.5 Salinity section through the Strait of Gibraltar from North Africa to Spain. The high salinity water at the bottom of the strait is Mediterranean outflow water flowing towards the Atlantic underneath a layer of less salty Atlantic water flowing into the Mediterranean Sea.

Calculating flows in the Strait of Gibraltar using the continuity principle

The equations used to calculate flow rates by specifying conservation of salt and water are known as Knudsen's equations after the Danish physicist Martin Knudsen. To see how they work, suppose the volume flow rate into the Mediterranean is F_{IN} (units m³/s), and let the volume flow rate out be F_{OUT} m³/s. Then the difference between these two must account for the loss by evaporation. The flow in must be a little bigger than the flow out. In fact, since the loss by evaporation, as we have seen, is 0.1 Sv,

$$F_{IN} - F_{OUT} = 0.1\,\text{Sv} \quad \text{for a water balance}$$

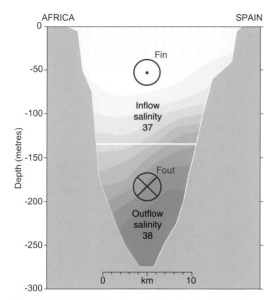

AFRICA   **SPAIN**

Fin

Inflow
salinity
37

Fout

Outflow
salinity
38

Depth (metres)

0   km   10

Figure 4.6 Sketch of the flows in the strait of Gibraltar. Atlantic water, of salinity about 37, is flowing towards you at the surface (circle with dot at centre). Mediterranean water, with salinity about 38, is flowing away from you (circle with cross).

A flow of F_{IN} m³/s of linity 37 will carry salt at a rate $37 \times F_{IN}$ kg s⁻¹ (since the salinity is the number of kg of salt per cubic metre of water). To maintain the salt balance, we would like the difference between the salt flowing in and that flowing out to be zero. That is:

$$37 \times F_{IN} - 38 \times F_{OUT} = 0 \qquad \text{for a salt balance}$$

These are two simultaneous equations for F_{IN} and F_{OUT}. The solution is obtained by dividing the second equation throughout by 37 and subtracting it from the first equation. This gives:

$$F_{OUT} = 3.7 \, Sv$$

$$F_{IN} = 3.6 \, Sv$$

38 times greater. Because the cross-sectional area of the flow is also now only half the cross section of the strait instead of the full cross section (the outflow occupies the other half), the speed of the flow is over 70 times greater, or nearly 3 m s⁻¹, or 6 knots. This is a considerable speed, and explains why sailing ships needed a fair wind to sail out of the Mediterranean against such a current.

In looking at Figure 4.5 you will notice that the interface between the inflowing Atlantic water and the outflowing Mediterranean water is not horizontal but slopes up towards Spain. The reason for this is that the flows are affected by the *spin of the earth*, which produces a phenomenon called the **Coriolis effect** (*see* next section).

4.4 The Coriolis effect

Soon after the invention of the barometer, it was discovered that winds, rather surprisingly, did not blow directly from high to low atmospheric pressure, but tended to follow isobars (i.e. blow along lines of constant pressure). The rule is that if you face into the wind in the northern hemisphere, the high pressure is to your left and low pressure to your right. This behaviour arises because we live on a spinning earth, and is called the Coriolis effect. The wind starts to blow from high to low pressure but is deflected to the right (in the northern hemisphere) because of the earth's spin. Moving objects in the southern hemisphere are deflected to the left. Sometimes we imagine that there is a force – the Coriolis force – acting to the right of moving objects in the northern hemisphere, and to the left in the southern.

The Coriolis effect occurs because the surface of the earth is moving at different speeds at different latitudes. Perhaps the easiest way to envisage it is to imagine you are looking down on the earth from above the North Pole. The earth will be turning below you in an anticlockwise sense. A point on the Equator is moving eastwards at a little over 1610 km h⁻¹. A point at 60° N (the latitude of the Shetland Isles) is moving eastwards at 805 km h⁻¹. A point at the North Pole is not moving at all: a person standing there would simply turn to face different directions as the earth turns, but would have no velocity.

Now imagine you travel from the Equator towards the North Pole. Because you start from the Equator, you will have a speed of over 1610 km h⁻¹ towards the east. Let's say there is not much friction between you and the solid earth below you, so you keep this initial speed as you move around. This idea of little friction with the solid earth is probably true of surface ocean currents or winds in the higher parts of the atmosphere. As you travel northwards you will move over places that are not moving as fast to the east as you are. *So, compared to the solid earth, you will appear to be deflected towards the east.* By the time you arrive at the Shetlands, you will be travelling at 805 km h⁻¹ towards the east from the point of view of someone standing on the Shetland Islands.

Ocean currents feel the Coriolis effect and are deflected by the Earth's spin. Although the account we have given is for a current moving northwards, the same is true for currents moving in any direction; they are always deflected towards the right in the northern hemisphere and to the left in the southern hemisphere.

Many ocean currents and winds are driven by **pressure differences**, that is they flow (or, in the case of winds, blow) from high pressure

to low pressure. As they do so, they turn to the right (in the northern hemisphere) because of the Earth's spin. As a result they never actually get to the low pressure, but start circling round the high pressure. This effect is most familiar in the atmosphere, where winds circulate around a high pressure in a clockwise sense in the northern hemisphere. We say that the wind is in **geostrophic balance**: as it travels around the high pressure, the pressure force acts outwards (towards the left, looking downwind) and the Coriolis force acts inwards (to the right, looking downwind). The two forces are in balance, and the wind can blow in steady state, that is at constant speed, with a zero net force. Notice that zero force does not mean zero motion. According to Newton's Laws, forces are required to produce *accelerations*. If there is no force there is no acceleration, but constant speed is perfectly acceptable. The geostrophic balance ignores frictional forces (*see* Figure 4.7).

If water is confined to flow along a sea strait, the Coriolis effect tends to pile the water up against one shore – the shore to the right looking down the direction of flow. The flows in the Strait of Gibraltar illustrate this effect nicely. The Atlantic water flowing into the Mediterranean experiences a Coriolis force to the right. This tends to pile up Atlantic water against the right hand shore (the African coast) creating a high pressure on this shore. There is then a pressure gradient force in the surface layer acting from the African shore towards Europe, and this just balances the Coriolis force acting in the opposite direction. To provide this pressure gradient force, the sea surface slopes up towards the African coast, but the slope is quite gentle, and cannot be seen with the eye. In the lower layer, the flow

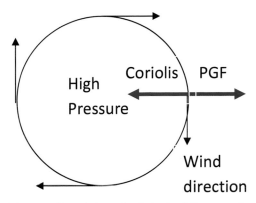

Figure 4.7 Force balance in winds circulating around a high pressure system in the northern hemisphere. The winds blow around the high pressure in a clockwise sense. The pressure gradient force (PGF) acts outwards, from high to low pressure, and is balanced by the Coriolis force acting towards the right of the wind direction. The balance between pressure gradient and Earth rotation effects is important in both meteorology and oceanography and is called a geostrophic balance.

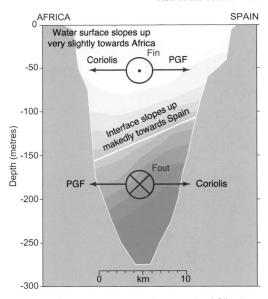

Figure 4.8 Geostrophic balance in the strait of Gibraltar. The Coriolis force on the outflowing Mediterranean water acts towards the right and pushes the Mediterranean water towards the Spanish side of the strait. This makes the interface slope up towards this side. The resulting pressure gradient force (PGF) then balances the Coriolis force in the bottom layer. In the upper layer, the Coriolis force pushes the flow towards the African shore, producing a sloping sea surface and a pressure gradient force acting from Africa towards Spain. The slope in the sea surface required to balance the Coriolis force in the upper layer is very gentle (of order 1 in 100,000) but the slope in the interface needs to be much greater than this because of the small density difference between the two layers.

is in the opposite direction, the Coriolis force is again to the right, which is towards Europe. This piles outflowing Mediterranean water against the European shore, creating a high pressure in the bottom layer on this shore. The interface between the two layers therefore slopes up towards Europe. This slope is quite considerable and can be observed by measurements of salinity at different depths. The flows and forces in a cross section of the Strait of Gibraltar are shown in Figure 4.8.

As the Mediterranean water leaves the Strait of Gibraltar and enters the Atlantic Ocean, it cascades down the sloping seabed, flowing through colder and fresher Atlantic water, which is less dense than the Mediterranean water. As the Mediterranean outflow water tumbles down the sea floor, it mixes with the overlying Atlantic water, becoming less salty

and therefore less dense. When it reaches a depth of about 1000 m, it reaches Atlantic water that is denser than itself, and it leaves the ocean floor and spreads out as a layer in the Atlantic. This layer of warm, salty water from the Mediterranean, at a depth of about 1000 m, can be detected throughout the north Atlantic, at great distances from the Strait of Gibraltar (*see* Chapter 1).

5 The tides

We give the name the *tide* to the regular rise and fall of sea level caused by the gravitational pull on the ocean by the Moon and the Sun. The accurate prediction of tides, in the form of tables, is one of the great success stories of marine science. Knowing about the timing and level of high and low waters can be crucial for the safe navigation of large ships in and out of port and through shallow sea straits. The vertical movement of the sea is driven by horizontal currents, called tidal streams. There is increasing interest in extracting the energy contained in tidal streams and using it to provide electricity for industry and homes.

5.1 How the tide is formed

The tide is produced by small differences in the Moon's (and to a lesser extent, the Sun's) gravitational pull over the surface of the Earth.

The Earth and Moon orbit once a month about their common centre of gravity, the **barycentre**. This orbiting motion produces centrifugal forces on the Earth analogous to those that cause a dancer's pony tail to fly out as they are whirled around by their partner. The motion of the Earth about the barycentre is such that each point on the Earth goes around the same circular orbit in the same time. The centrifugal force is therefore the same at all points on, and in, the Earth (dashed arrows in Figure 5.1). Now, these centrifugal forces are balanced *on average* by the Moon's gravitational pull and this stops the Earth and Moon flying apart. In fact, the Moon's gravitational pull exactly balances the centrifugal force at the centre of the Earth and this is sufficient to keep the solid body of the Earth in the correct motion about the barycentre. But, as with all heavenly bodies,

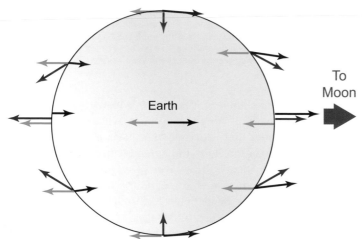

Earth

To Moon

Figure 5.1 The origin of the Tide Generating Force. Centrifugal forces arising from the Earth's orbital motion about the common centre of gravity of the Earth and Moon (the barycentre) are shown as blue arrows. These point away from the Moon and are the same size at all points on and within the Earth. The Moon's gravitational attraction is shown as black arrows. This exactly balances the centrifugal force at the centre of the Earth. At most other points there is a small imbalance between the Moon's gravitational pull and the centrifugal force. It is this small imbalance which is the tide generating force. In this figure, this force is shown as red arrows, which have been exaggerated in size for clarity.

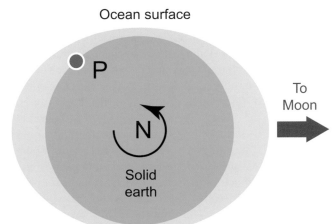

Ocean surface

P

N

Solid
earth

To
Moon

Figure 5.2 If the Earth were covered in an ocean, the tide generating forces would tend to pull the ocean into the shape of a rugby ball, with points facing towards and away from the Moon. In this view, we are looking at the Earth from above the North Pole. As the Earth turns, a point P will experience two high waters each day, one when the Moon is directly overhead, the other when the Moon is directly under foot.

the gravitational pull of the Moon diminishes inversely as the square of its distance. The Moon's gravitational pull therefore gets less as you move over the surface of the Earth in a direction away from the Moon (solid black arrows in Figure 5.1). This produces local imbalances between the Moon's gravity and the centrifugal forces, and these imbalances are represented by the orange arrows in Figure 5.1. It is these small differences between the Moon's pull and the centrifugal forces (called the **tide generating forces)** that create the tide in the ocean.

If the Earth were covered in an ocean, it is possible to imagine that the red forces shown in Figure 5.1 would pull the ocean into a bulge directly below the Moon, and another bulge on the opposite side of the Earth to the Moon. That is, the ocean will take on the shape of a rugby ball with the points in line with the Moon (Figure 5.2). As the Earth rotates within this ocean, a high tide will be experienced when the Moon is most directly overhead (called the Moon's transit) and another when the Moon is most directly under foot. It is possible to

calculate the size of the bulges expected from gravitational theory – they are about 0.5 m. So the sea level will go up and down by about 0.5 m twice a day. In fact, the time for two high tides will be slightly longer than a day because, in the time it takes for the Earth to turn once, the Moon moves a little in its orbit and the Earth has to rotate a little more to catch up with this movement. In 24 hours, the Moon moves about 1/30th of the way around its orbit and it takes 1/30th of a day – about 50 minutes – for the Earth to catch up. The time for two complete high waters is therefore 24 hours and 50 minutes. High waters are therefore separated by 12 hours and 25 minutes.

The account given above is a summary of the *equilibrium theory* of the tide proposed by the English scientist Isaac Newton. In addition to the tides produced by the Moon, the motion of the Earth about the Sun also produces tidal forces. These are somewhat weaker (by a factor of 0.46) than those of the Moon but they produce important modulations in the strength of the tide generating force. When the Sun, Earth and Moon are in line (at new

The tides of Europa

The tide generating force depends on the mass and distance of the object which is producing it. It can be shown that the force varies inversely as the *cube* of the distance to the object. Note that this is different to gravitational forces which vary inversely as the square of the distance. Tidal forces represent the rate of change of gravity with distance and change with (distance)$^{-3}$. For this reason, tidal forces increase rapidly as the distance to the object shortens. Tide generating forces are therefore particularly large on moons orbiting close to large planets such as Jupiter and Saturn. Europa is the sixth moon of Jupiter and a little smaller than our own moon. It is covered in ice, but evidence has been accumulating for a number of years that below the surface of the ice there is an ocean of salt water. The effect of tidal forces grinding the ice and, through the friction generated in this way, warming the interior may be enough to maintain liquid water on this otherwise frozen moon. There is also evidence that cracks may form in the surface of the ice, allowing sunlight into the liquid water. If so, that could drive photosynthesis. The tidal forces on Europa therefore offer one of the best chances in the solar system of extra-terrestrial life.

and full moon) the tide generating forces are at their maximum. When the Sun and Moon make an angle of 90° with the Earth, the tide generating forces are at their minimum.

Newton was aware of the weaknesses of his equilibrium theory. It requires an earth which is covered in an ocean, which is clearly not the case. Also it requires the Earth to turn slowly enough that it does not disrupt the tidal bulges. This is also not the case. The Earth spins too fast for the tidal bulges to retain their position in line with the Moon; we return to this point in section 5.3. Nevertheless, the equilibrium theory explains several of the most important features of the observed tide. This will be examined in the following section, followed by a discussion of how the discrepancies between the equilibrium theory and the real tide can be explained by considering the motion of the tide as a wave.

5.2 Observing the tide

The tide in the Earth's oceans and seas can be measured directly by recording the level of the sea on a vertical surface such as that of a pole driven into the seabed or a harbour wall. Alternatively, a pressure sensor at a fixed point can be used to record the changes in pressure due to the rise and fall of sea level. Figure 5.3 shows a short series of measurements at Menai Bridge in North Wales. In this 24 hour period, there are two high tides and two low tides, the high tides separated by just over 12 hours. Locations where there are two high tides each day, which is the common situation, are said to experience a **semi-diurnal tide**.

Like most places, high water at Menai Bridge does not occur at the time of the Moon's transit (that is the time the Moon is due south). In fact, here high water occurs about 10 hours after the Moon's transit. The water surface elevation is measured relative to a fixed level or datum. In Figure 5.3, the elevation is

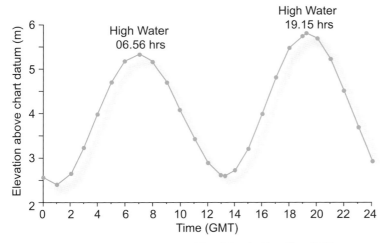

Figure 5.3 The tide at Menai Bridge in North Wales during a single day, showing the two high and two low waters in 24 hours that are characteristic of a semi-diurnal tide.

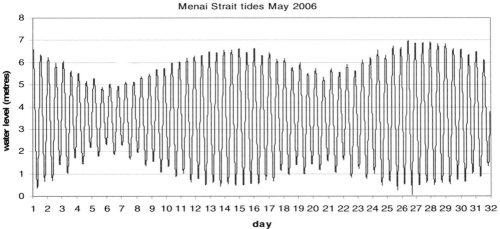

Figure 5.4 The tide at Menai Bridge over a period of a month, showing the changes in tidal range associated with spring and neap tides. Spring tides occur on the 14th and 27th of the month and neap tides on 6th and 20th.

expressed relative to **chart datum**. This is the lowest level that the water is likely to reach at this location in normal weather conditions. As the name implies, depths on nautical charts are expressed as the depth below chart datum; this means that the depths on charts are the least possible: normally, the tide will increase the depth above that shown on the chart.

The **tidal range** is the vertical distance between the lowest water level (or low water) and the high water in a tidal cycle. In Figure 5.3, the tidal range is about 3 m. At a given location, the tidal range changes from day to day in a fortnightly cycle. Figure 5.4 shows the variation in water level at Menai Bridge over a month. The periods of largest tidal range are called **spring tides** and they occur at or shortly after the times of new or full moon. The times of lowest tidal range, called **neap tides**, occur at about the time of half moon. Notice that spring tides have the highest high waters and also the lowest low waters. The word 'spring' in the tidal context comes from the same Anglo-Saxon root as a spring in the ground, meaning an upwelling of water, rather than referring to the spring season.

The tidal range at a coastal location also varies with other rhythms. It is noticeable in Figure 5.4, for instance, that the range at the two periods of neap tides is not the same: the second neap tide in the month has a greater range than the first. This difference is caused by changes in the Earth–Moon distance. The lunar orbit is elliptical and when the Moon is closest to the Earth (perigee) the tide gener-ating forces are greater than when it is at its furthest from Earth (apogee). The semi-diur-nal tide is greatest when the Moon and Sun are in the plane of the Equator. In an extreme case, if the Moon were above the North Pole, it would produce no tide in the ocean at all. For this reason, the largest semi-diurnal tides of the year occur at spring tides at the equi-noxes in March and September, particularly if the Moon is also in the plane of the Equator. Exceptionally large semi-diurnal tides occur when the Moon is at perigee and when both Moon and Sun lie in the plane of the Equator. This rarely happens exactly, but there are more frequent occasions when these conditions are nearly met. We are due to have exceptionally large tides next in March and September 2015.

The angle the Sun and Moon make with the plane of the Equator is called the solar (or lunar) declination. When the lunar declina-tion is not zero, there are differences in the height of the two high waters in the day. This can be seen to some extent towards the end of the record in Figure 5.4. The effect is called the diurnal inequality. The diurnal inequality is not great on the coasts of northern Europe, but can be marked in other parts of the world, notably the west coast of the United States and the north coast of Australia. In a rela-tively small number of locations, the diurnal inequality can grow to the extent that there is effectively just one high water and one low water in each day. Such locations are said to have a **diurnal tide**. The Gulf of Carpenteria on the north coast of Australia is an example of a place with diurnal tides.

5.3 The tide as a wave

In the picture described above, we have imagined that the bulges in the ocean shown in Figure 5.2 stay aligned with the Moon as the Earth turns. In fact the ocean bulges shown in this figure correspond to the crests of a wave stretched right around the circumfer-ence of the Earth. If the crests are to stay in line with the Moon, this wave must travel over the surface of the Earth at the same speed as the Earth rotates. At the Equator, the speed of Earth rotation is over 1610 km h^{-1} and waves in the ocean cannot travel this fast. Further-more, the presence of continents will stop the continuous movement of waves right around the world everywhere except in the southern ocean.

As the Earth turns, the tide generating force pushes the water in the ocean first one way then the other, each 'push' lasting for a little over 6 hours. It is actually only the horizon-tal component of the force that is important in this regard; the vertical component can be neglected in comparison to the Earth's own

Ocean surface

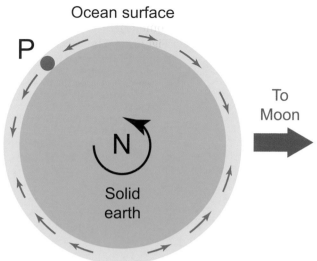

P

To
Moon

N

Solid
earth

Figure 5.5 Illustrating the horizontal component of the tide generating force, called the tractive force, as brown arrows. The view of the Earth is from above the North Pole with the Moon over the Equator. As the Earth turns, the tractive force pushes the water of the ocean first one way and then the other, creating a wave.

gravity – *see* Figure 5.5. This regular pushing creates waves in the ocean that have the same period as that of the tidal forcing – 12 hours and 25 minutes (also known as 12 **lunar hours**) for the semi-diurnal tide. The speed of the waves depends on the depth of water in which they are travelling. The wavelength is long compared to the depth of the ocean, and so they travel at a speed equal to √(gD) where g is the acceleration due to gravity and D the water depth.

The tidal waves in the ocean are produced directly by the tide generating force. Shelf seas are generally too small to feel the effect of the tidal forces directly. Instead, the rise and fall of the ocean sends waves (with tidal period) travelling from the shelf edge towards the coast. The speed and wavelength of these waves just depends on the water depth. For example, in a shelf sea of depth 100 m, the speed of the wave is 113 km h^{-1} and the wavelength over 1400 km.

Figure 5.6 shows such a tidal wave progressing along a coastline. High water occurs when the crest of the wave arrives at a point on the coast. In these long progressive waves, the currents flow with the wave underneath the crest and against the direction of wave travel under the trough. The effect of earth rotation is to deflect the currents to the right in the northern hemisphere. The water is piled up on the right hand shore of the canal at the crest and at the left hand shore at the trough (Figure 5.7). This piling up then creates a pressure gradient that acts to balance the Coriolis force. The wave is in geostrophic balance, as we discussed in Chapter 4. The overall effect on the wave is that it now has a greater amplitude on the right shore of the canal than on the left, as we look down the canal in the direction the wave is travelling. Such a wave is called a **Kelvin wave** named after the Scottish physicist, Lord Kelvin.

You might wonder why wind waves in a regular-sized canal (such as used by narrow boats) don't have a greater amplitude on one side of the canal. The reason is that the

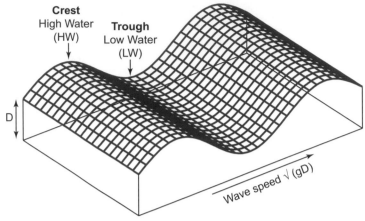

Crest
High Water
(HW)

Trough
Low Water
(LW)

D

Wave speed √(gD)

Figure 5.6 Sketch illustrating a progressive tidal wave. High water (HW) is at the crest of the wave and low water (LW) at the trough. The crest of the wave is travelling from left to right at the wave phase speed $\sqrt{(gD)}$ where g is the acceleration due to gravity and D the water depth. Maximum currents flow in the direction of wave travel at high water and in the opposite direction at low water.

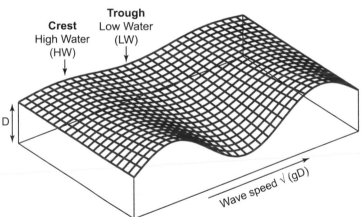

Crest
High Water
(HW)

Trough
Low Water
(LW)

D

Wave speed √(gD)

Figure 5.7 A Kelvin wave in the northern hemisphere. The wave is travelling from left to right. Beneath the crest, where the water is flowing with the wave, the Coriolis effect pushes water towards the right and the height of the high water increases in this direction. Below the trough, the current is flowing against the direction of wave travel. The Coriolis effect pushes water to the right of the current and the height of the low tide increases in this direction. The result is a wave that has a greater tidal range on the right hand shore looking down the direction of travel.

Coriolis effect is really only observable in motions that last long enough for the effect of Earth rotation to matter. Waves in a narrow-boat canal have a period of a few seconds, and the Earth doesn't turn very far in a few seconds. Semi-diurnal tidal waves have a period of about 12 hours, and they respond to the effect of the Earth's spin.

5.4 Co-tidal charts

A co-tidal chart is a good way of showing the wave-like behaviour of the tide. On it are plotted *co-tidal* lines, which join places where high water occurs at the same time, and *co-range* lines, which join places that have the same tidal range. In the case of a progressive wave travelling down a channel, for instance, the co-tidal lines will be a series of parallels, marking the progressively later time of arrival of the high water with distance down the channel. In places where the wave is travelling fast the lines will be well spaced, and they will be close together where the wave slows down. If the wave is a Kelvin wave, the co-range lines will show a reduction of tidal range from one shore of the channel to the other.

As an illustration, Figure 5.8 shows a co-tidal chart for the semi-diurnal tide in the Irish Sea. A tidal wave enters the Irish Sea through its southern entrance and travels up the channel between Wales and Ireland. The tide here behaves as a progressive wave, taking four hours to travel from the south to the north of Wales. The effect of Earth rotation on the

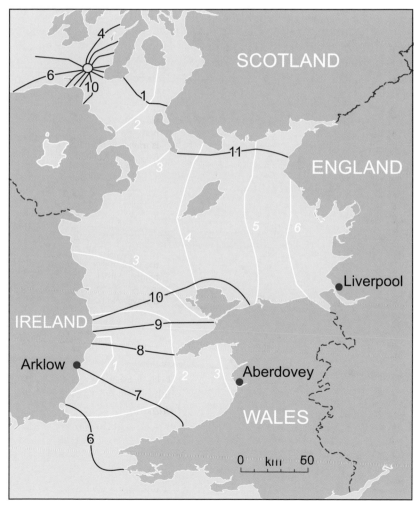

Figure 5.8 Co-tidal chart for the main lunar semi-diurnal tide in the Irish Sea. Along each co-tidal line (shown as continuous lines) high water occurs at the same time. Times marked on co-tidal lines show the time of high water in hours after the Moon's transit at Greenwich. Along each co-range line (shown as dashed lines) the tidal range is the same. Ranges are marked in metres.

wave can clearly be seen: i.e. the tidal range at Aberdovey on the Welsh coast is more than three times greater than that at Arklow at the opposite place on the Irish coast.

In the northern part of the Irish Sea, the co-tidal lines disappear and the tidal range continues to increase towards Liverpool and the English coast. High water occurs throughout much of the northern Irish Sea at the same time (give or take half an hour). The reason why this happens is that the progressive wave travelling up between Wales and Ireland reflects off the English coast and bounces back into the Irish Sea. The tides in the northern Irish Sea are then the sum of the incoming and reflected waves. When two waves travelling in opposite directions add together, they create a type of wave called a **standing wave**.

5.5 Standing waves

Consider a wave travelling into the gulf shown in Figure 5.9 and being reflected without loss of energy at the head of the gulf. The gulf will then contain two equal-sized waves travelling in opposite directions, and the shape of the water surface (and the currents) at any time will be the sum of that in the two waves. At a location one quarter of a wavelength from the head of the gulf a **nodal line** is formed. Along this line the vertical motion of the two waves always

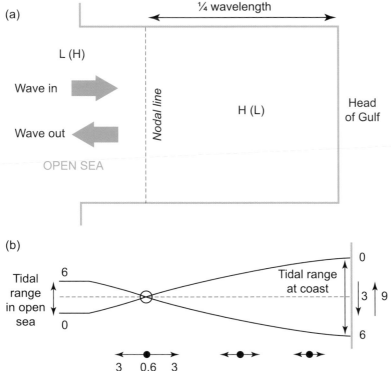

Figure 5.9 Illustration of the formation of a standing wave in a gulf connected to the open sea: (**A**) view from above; (**B**) side view. A wave enters the gulf from the open sea, is reflected from the head of the gulf and travels back out again. The two waves add to form a standing wave with a nodal line (along which there is no tide) located one quarter wavelength from the head of the gulf. On one side of the nodal line it is high water when it is low water on the other side and vice versa. Considerable amplification of the tidal range at the coast can result if the nodal line is located close to the boundary between the gulf and the open sea. The numbers 0, 3, 6, 9 in the lower figure refer to hours during the tidal cycle

cancels each other – it is high tide for one wave when it is low tide for the other and vice versa. There is therefore no vertical tide along the nodal line. On one side of the nodal line it is low water when it is high water on the other side, and vice versa. Moving away from the nodal line, the tidal range increases (Figure 5.9B).

The currents in the two waves also add at the nodal line, but in this case they reinforce each other so that the nodal line is a place of fast currents. The maximum speed of the currents decreases with distance from the nodal line. The cycle then proceeds as follows. At lunar hour 0 (let's say) it is high tide at the head of the gulf and the current is everywhere zero. At hour 3, the water is flowing at its fastest away from the head of the gulf through the nodal line and the water level is falling at the head of the gulf. At hour 6 it is low water at the head of the gulf and the currents are slack again. At hour 9, the current flows at its fastest towards the head of the gulf and the water level there is rising. The motion is analogous to the sloshing wave that can be created in a bath by moving a hand back and forth. It is high water at one end of the bath when it is low water at the other and vice versa. In between, when the water level is the same at the two ends, the water is flowing at its fastest from one end to the other.

Where the gulf joins the ocean, the tidal range must match that of the ocean tide (Figure 5.9B). If the nodal line lies close to this junction, it means that the tidal range at the head of the gulf will be much greater than that in the ocean. This effect amplifies the tide in coastal seas, so that a small tidal range in the ocean can become a much larger range at the coast. The sea is said to be in **resonance** with the applied forcing: a small force, applied regularly at just the right time, produces a large response. The equivalent in the bathtub is that, if the forcing is timed just right, the water can be sloshed out of the bath.

In order to achieve a large amplification in the illustration of Figure 5.9, the gulf should be close to ¼ wavelength from head to mouth. Since the wavelength of the tide in the gulf just depends on the depth of water, resonance will occur for particular values of water depth and gulf length (or width of shelf sea from shelf edge to coast). Places with large tidal ranges, such as the Bay of Fundy in Canada, have these critical values. We have drawn Figure 5.9 for a gulf which is just over ¼ wavelength long, but resonance can also occur in gulfs which have a length equivalent to ¾ or 1¼ tidal wavelengths. The North Sea, for example, is a gulf which is approximately 1¼ tidal wavelengths long, and which has 3 nodal lines (at ¼, ¾ and 5/4 wavelengths from the coast).

5.6 Amphidromic systems

On a turning earth, the currents in the tidal wave respond to the Coriolis effect. When the water is flowing out of the gulf (at hour 3 in Figure 5.10) the Coriolis effect creates a slope in the water surface along the nodal line, so that it is no longer a line of no tide. The slope creates a pressure gradient that balances the Coriolis force, creating a geostrophic balance. The slope is upwards towards the right hand shore looking from the head of the gulf. It is therefore high tide on this shore (and low tide on the left hand shore) at hour 3. When the current is flowing into the gulf (at hour 9) the Coriolis effect reverses the slope so that it is high tide on the left hand shore at hour 9. The water surface along the nodal line slopes one way and then the other, like a see-saw. Only the

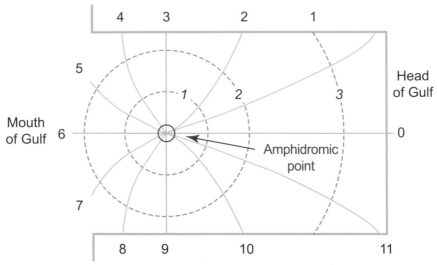

Figure 5.10 Co-tidal chart for an amphidromic system in the northern hemisphere. Co-tidal lines are continuous, with times of high water marked in lunar hours (high water at the head of the gulf occurs at hour 0). Dashed lines are co-range lines. Earth rotation effects turn the no-tide nodal line into a single no-tide amphidromic point about which the tidal wave turns in an anticlockwise sense. The tidal ranges increase (and the tidal currents decrease) with distance from the amphidromic point.

central point of the nodal line remains a place with no tide (it is equivalent to the fulcrum of the see-saw). The point of no tide is called an **amphidromic point**. The tidal wave appears to swirl around this point in an anticlockwise direction in the northern hemisphere, rather like the way that you can swirl a wave around the edge of a teacup. The wave sweeping anticlockwise around the nodal point is called an *amphidromic system*. The tidal range increases (and the tidal currents decrease) away from the amphidromic point.

Friction between the tidal currents and the seabed reduces the energy in the tidal waves, and this means that the wave height (or in this case tidal range) becomes less as it travels along. The effect of friction on an amphidromic system is to move the amphidromic point along the nodal line towards the left-hand

shore looking into the gulf. In the southern Irish Sea, frictional effects are so great that the reflected wave has decayed away to virtually nothing, and the tide here behaves essentially as an incoming Kelvin wave modified to some extent by a weak reflected wave. The modification can be seen in the fact that the co-tidal lines are converging on a point that lies on dry land in Ireland. We can imagine that the amphidromic point has been moved so far to the left that it is no longer within the sea (it is called a degenerate amphidrome). A second tidal wave enters the Irish Sea through the North Channel between Ireland and Scotland. This reflects off the English coast and forms a standing wave in the northern half of the Irish Sea. Frictional effects are apparently not so great in the case of this wave, and an amphidromic point is formed in the North Channel.

5.7 Predicting tides

Tide tables, showing the times and heights of high and low water for important coastal towns, appear in daily newspapers and on the internet and are a vital aid for coastal navigation. The science of predicting tides is based on a curve fitting technique. Observations of water level are made at a port for a length of time: this needs to be a year or more to produce the most accurate tide tables. A number of sinusoidal curves with a fixed period – called tidal harmonics – are then fitted to this data. The period of these harmonics is set by the motion of the Earth, Moon and Sun. For example, an important harmonic, called the main lunar semi-diurnal harmonic, has a period of 12 hours and 25 minutes. Other harmonics express the Sun's tide and the changes in the distance and declination of the Moon and Sun. The amplitude and phase of these harmonics is adjusted mathematically so that when they are added up they give the best possible fit to the observations. These best-fit amplitudes and phases are fixed for a given place and are called the tidal constants. Once they have been determined, each of the harmonics can be plotted for any day in the future; they can be added together and the times and levels of high and low tide determined.

Tide tables are available for many places, but not everywhere. If you want to make your own tide tables for one of the places not covered, you can do so by determining the time difference between high water at your chosen location and that at a nearby port where tide tables are available. This tidal correction should be fairly constant as long as the port is not too far away. You can also apply a correction to the tidal range.

The modern method of tidal prediction requires mathematical analysis, which is easily done on modern computers. Historically, accurate tidal tables were produced for important ports such as Liverpool and Bristol long before computers were available. The makers of these tables (for example, the Holden brothers in the case of Liverpool) refused to reveal their methods, taking their secrets to the grave. It seems likely, however, that their method was based on the equilibrium tide with corrections to the time and height of high tide to bring them into line with observations.

5.8 Tidal streams and tidal energy

The horizontal movements of water associated with the tide are called tidal currents or tidal streams. Near the shore, or in an estuary, the flow of water towards the land is called the flood, and the flow of water away from the land the ebb. Tidal streams can be very fast, up to several knots, but they don't flow for very long in a given direction. In a symmetrical tide, the flood and ebb will each last for 6 lunar hours, or 6 hours and 12 minutes.

There is currently great interest in developed countries in extracting some of the energy contained in tidal streams. The best places to do this are where the streams are at their fastest. An amphidromic point therefore makes a good place to put tidal stream power plants. Tidal streams are also influenced by local geography. For example, a narrow channel can have fast streams and they can also be fast close to a headland.

Turbines for generating usable energy from tidal streams need to be able to cope with the changing direction over a tidal cycle. The power that can be extracted depends on the cube of the tidal current speed, and so rises rapidly as the speed increases. Tidal stream energy is reliable and predictable. By placing a series of turbine farms at different places around the coast, peak generation can be achieved at different times of day to meet local demands.

Alternatively, tidal power can be extracted by using dams or barrages. The flood tide is allowed to freely enter a basin behind the barrage and then the water drains out through electricity-generating turbines when it is low tide in the open sea. The principal objection to tidal barrages, apart from high cost, is that they raise mean sea level within the barrage, thus flooding tidal flats that are important feeding grounds for birds.

6 Stratification and fronts in shelf seas

Shelf seas in temperate latitudes undergo seasonal heating and cooling. We shall see in this chapter that this has an important effect on the structure of the water column, creating layering of the water in the summer. This, in turn, has implications for the biology of shelf seas, which are explored in Chapter 10.

6.1 The seasonal thermocline

On a summer's day in a deep part of a temperate shelf sea where the tidal currents are not too strong, observations show that the water has thermal structure: it is divided into layers, with a surface layer of warm water lying on top of a deeper layer of cooler water (Figure 6.1). Immediately below the surface is a layer that is warmed by the sun and stirred by the wind, called the **surface mixed layer**, or the wind-mixed layer. Below this, the temperature decreases with depth, in a layer called the **seasonal thermocline**. Below the thermocline in shelf seas, there is usually another mixed layer (the bottom mixed layer) extending to the seabed. When the water is layered in this way it is said to be *stratified*.

The seasonal thermocline forms in early spring (late March or early April in the northern hemisphere). It develops first in the deeper parts of the shelf where the tidal currents are weakest. Around the British Isles these areas are in the northern North Sea and the Celtic Sea south of Ireland. The stratified area rapidly spreads into shallower water in the next few weeks, and at the same time it intensifies. Most of the sun's heat now goes into the surface mixed layer, and so the temperature difference between surface and bottom of the water column increases. By mid-summer, the temperature difference between the surface and bottom can be several degrees, with most of this occurring in just a few metres through the thermocline. After the summer solstice, the sea surface starts to cool, and at the same time wind speed tends to pick up as autumn approaches. The surface mixed layer now deepens and the thermocline is driven down towards the sea floor. This process

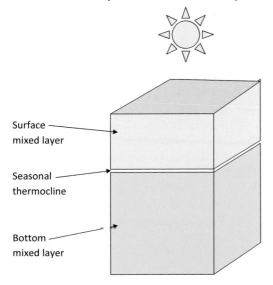

Surface mixed layer

Seasonal thermocline

Bottom mixed layer

Figure 6.1 Illustration of a stratified water column, common in shelf seas in summer. A sun-warmed and wind-stirred surface mixed layer is divided from deep, cool tide-stirred water by a seasonal thermocline.

transfers deep, cooler water into the surface mixed layer, lowering its temperature further. Eventually the water column becomes vertically mixed. This takes time, though, and the remnants of stratified water can linger in shelf seas in the northern hemisphere up until the beginning of December.

During winter, shelf seas cool, giving the summer warmth up to the atmosphere. The sinking of water cooled at the surface in *convective overturning* ensures that the water column remains vertically mixed until warming starts again in the following spring.

Even in summer, there will be parts of the shelf which remain immune to thermal stratification. These will include shallow water close to the coast (perhaps less than 10 m deep) where wind and waves together create enough turbulence to mix the sun's heat from surface to seabed. In deeper water, tidal currents can be very effective at vertical mixing, and so regions with very fast tidal streams, for example near amphidromes (*see* Chapter 5) can remain vertically mixed throughout the year.

6.2 Stability and mixing in the thermocline

When the seasonal thermocline is present in the summer months it acts as a barrier to the vertical transport of material in the sea. This has important consequences for biology and sediment transport in shelf seas. In the spring, phytoplankton 'bloom' (or grow rapidly) in the well-lit surface mixed layer (more details of this are given in Chapter 10). This spring bloom strips nutrient salts out of the surface layer, but not out of the bottom layer, which is too dark for phytoplankton to grow. As a result, in the summer, the thermocline divides a well-lit, but nutrient-depleted surface layer from a dark, but nutrient-rich bottom layer.

The reason why the thermocline acts as a barrier to vertical mixing can be understood if we remember the effect of temperature on water density (Chapter 2). Warm water is less dense than cold water. If we try and push some water from the surface mixed layer down through the thermocline it will be pushed back by its buoyancy (if you've ever tried to push a beach ball into the sea you will be familiar with this effect). Similarly, if we try to lift some water from the bottom mixed layer into the surface mixed layer, it will sink back down again because it is denser than the water into which it has been introduced (Figure 6.2). It is possible to mix water through the thermocline, but it takes energy to do so.

There are two principal sources of energy available for vertical mixing in shelf seas. These are provided by (a) the effects of the wind on the sea surface generating mixing in the surface mixed layer and (b) tidal streams rubbing against the sea floor generating mixing in the bottom mixed layer. It is these sources of energy that create the mixed layers, and they are also able to produce mixing between the layers (and hence across the thermocline) by a process called **entrainment** (section 6.4).

6.3 The thickness of the surface mixed layer

Figure 6.3 shows a useful way to envisage seasonal stratification in shelf seas. The surface mixed layer is warmed by the sun and stirred by the wind (depicted in this figure by a mechanical stirrer). The bottom mixed layer is stirred by the tide (depicted by another mechanical stirrer) and a thermocline separates the two layers. We can then use a simple, but rather elegant, energy model to show how the thickness of the surface mixed layer depends on the wind speed and the rate of warming by the sun.

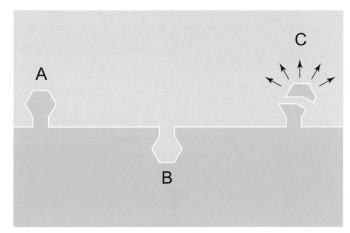

Figure 6.2 The thermocline barrier. Warm surface water is coloured pink and cold bottom water blue. An intrusion of cold water into the upper warm layer, as at **A**, will sink back down again because it is denser than the surrounding warm water. Similarly, an intrusion of warm water downwards (at **B**) will rise back because it is buoyant. It is only if the intrusion is given enough energy to overcome the density difference that transfer of water across the thermocline can take place. In the example at **C**, cold bottom water is being entrained into the surface layer.

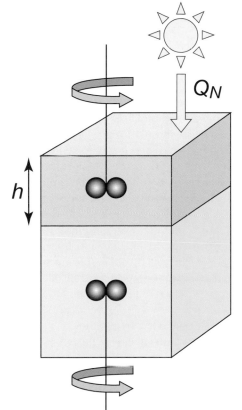

Figure 6.3 Heating–stirring model of the seasonal thermocline. The sea is heated by the sun (at a net rate Q_N) and stirred at the surface by the wind and at the bottom by the tide (stirring is represented here by turning paddles). Wind stirring drives the thermocline down and tidal stirring drives it up. The resulting position of the thermocline (and the thickness of the layers) depends on the balance between these two processes.

As the surface layer is warmed by the sun it expands, and therefore its **potential energy** increases as its centre of gravity rises. It can be shown that the rate of increase of the potential energy of the surface mixed layer is proportional to its thickness times the net rate of heating (that is the solar flux of heat through the surface minus the heat losses back to the atmosphere). If we call the thickness of the surface mixed layer h and the net rate of heating Q_N, then the rate of increase of potential energy is proportional to the product hQ_N. Now, the rate at which the wind puts energy into the sea is proportional to the cube of the wind speed (this is because the friction force of the wind on the sea depends on the square of the wind speed, and the rate of work done by this force is equal to the force times the wind velocity). Most of the energy put into the sea by the wind is dissipated as heat or used to make waves, but a small fraction (less than 1 %) is utilised in mixing the sun's heat down and creating the surface mixed layer. Let's suppose that a constant fraction of the wind energy is used for this purpose. Then the energy available from the wind to create the surface mixed layer is proportional to w^3 where w is the wind speed. In steady state, the energy input just balances the energy increase and something that is proportional to w^3 is equal to something that is proportional to hQ_N. That is

$$h = C\frac{w^3}{Q_N}$$

where C is a constant. Experiments show that if h is expressed in metres, w in $m\,s^{-1}$ and Q_N in Wm^{-2} then C is about equal to 8.

Equation 6.1 can be used to estimate the thickness of the surface mixed layer if the wind speed and heat flux through the sea surface are known. The latter can be estimated from changes in temperature of the water. If the water is warming up, this implies a positive heat flux through the surface. If the water is cooling down, the sea is giving heat back to the atmosphere. Table 6.1 below shows estimates of Q_N, calculated in this way, for a fixed observatory station (station E1) in the English Channel, operated by Plymouth Marine Laboratory. Notice that the net heat flux is positive for six months (the sea is warming up) and negative for the other six months. The table also shows monthly mean wind speeds measured at Aberporth on the Welsh coast. The depth of the surface mixed layer, calculated from equation 6.1, is also shown.

Table 6.1: Estimates of the net rate of heating (Q_N), calculated for a fixed observatory station (station E1) in the English Channel operated by Plymouth Marine Laboratory. h is the height of the mixed layer depth and w is the wind speed.

Month	Q_N $W\,m^{-2}$	w $m\,s^{-1}$	h m
January	−57	7.4	–
February	−50	7.2	–
March	−14	6.9	–
April	50	6.1	36
May	99	5.6	14
June	92	4.9	10
July	78	4.7	11
August	48	5.6	29
September	5	5.9	330
October	−53	6.4	–
November	−110	7.1	–
December	−86	7.9	–

Notice that a mixed layer depth cannot be calculated when Q_N is negative. A surface mixed layer will only form when the surface is being heated. When it is being cooled, convection will overturn the water column and the mixed layer depth will be the same as the water depth. Notice also that for September, a surface mixed layer depth of 330 m is calculated. Since the total water depth here is only about 50 m, this means that actually the water column will be well mixed in this month, even though it is still being heated weakly at the surface.

6.4 Entrainment

The surface mixed layer deepens in the autumn months through the process called **entrainment**. The turbulence generated in the surface layer by the wind produces undulations (or internal waves, *see* Chapter 2) in the thermocline. If these become sufficiently vigorous, the tops can break off, transferring parcels of water from the lower into the upper layer. Entrainment always takes place from relatively still water towards more vigorously stirred water. The entrained water makes the volume of the surface layer increase (and that of the bottom layer decrease) and as a result the thermocline moves down at a speed called the entrainment velocity. The entrained water transfers material from the bottom layer into the surface layer. For example, it will transfer nutrients that are needed by phytoplankton (Chapter 10).

At the same time as entrainment into the surface layer is occurring, stirring in the bottom layer by tidal mixing entrains surface water into the bottom mixed layer. Since the surface water is warmer, this produces a warming of the bottom mixed layer, which can be observed in records of bottom water temperature (Figure 6.4).

It is possible to work out the entrainment velocity due to tide mixing from the rate of warming of the bottom layer. The entrainment velocity is equal to this rate of warming multiplied by the thickness of the bottom layer and divided by the temperature difference between the surface and bottom layers. The thickness of the bottom layer at station E1 in summer is about 30 m. The rate of warming of the bottom layer (from Figure 6.4) in summer is about 5 °C in 6 months or about 0.03 °C per day. In June, the temperature difference between top and bottom is about 2 °C. Putting these figures into the formula above gives an entrainment velocity of about 0.5 m per day. The thermocline will not necessarily move upwards at this speed because wind mixing is driving it downwards at the same time. The net rate of movement of the thermocline will be the difference between the upwards entrainment velocity due to the tide and the downward entrainment velocity due to the wind. It is possible to have a situation in which the thermocline is not moving, but water is being entrained across it at equal rates by wind and tidal stirring.

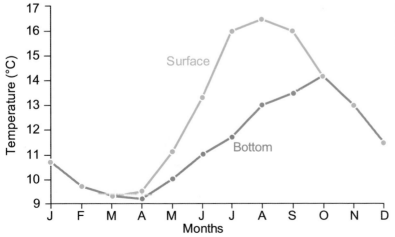

Figure 6.4 Annual variations in surface and bottom temperature at station E1 in the English Channel.

6.5 Tidal mixing fronts

If the combined stirring by wind and tide is sufficiently vigorous, it can prevent the seasonal thermocline forming altogether, and the water remains vertically mixed throughout the year. This happens in large parts of the north-west European shelf where the tidal currents are fast, including much of the Irish Sea, the English Channel and the southern North Sea. In other places, where the tidal streams are not so fast, a seasonal thermocline forms in the summer. The places where the transition occurs between vertically mixed and stratified water are called **tidal mixing fronts** (Figure 6.5).

On one side of the front, the water is thermally stratified, with warm water lying on top of cold. On the other side of the front, the water is vertically mixed, with uniform cool water from sea surface to the seabed. There are often changes in water colour and transparency, which are noticeable to viewers on a ship or aeroplane crossing the front (figure 6.6). There is also a difference in surface temperature. For this reason, tidal mixing fronts are visible from space in thermal infra-red imagery (Figure 6.7).

Unlike weather fronts in the atmosphere, tidal mixing fronts form in the same places each year. This is because their position is controlled by the geographical distribution of the strength of the tidal streams, and this doesn't change with time: places that have strong tidal currents *always* have strong tidal currents.

We can derive a criterion for the location of fronts using an extension of the energy argument of section 6.3. For water to be vertically mixed, there has to be a sufficient supply of stirring energy to provide the increase of potential energy due to thermal expansion. For a vertically mixed water column of depth h, the rate of increase of potential energy as it is heated at net rate Q_N is proportional to hQ_N. The rate at which energy is supplied by the wind is proportional to the cube of the wind speed. Similarly, the rate at which energy is supplied by the tide is proportional to the cube of the tidal current speed. Equating the rate of supply of stirring energy to the rate of increase of potential energy gives:

$$au^3 + bw^3 = hQ_N$$

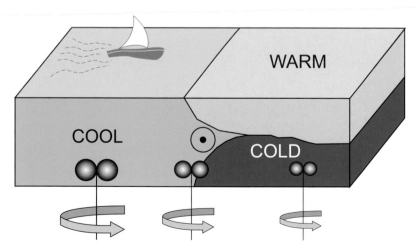

Figure 6.5 A tidal mixing front divides stratified and vertically mixed water in the summer in shelf seas. At the surface the front is characterised by a transition from warm to cool water. The turning paddles show the weakening gradient of tidal stirring in the direction of the stratified water. The dotted circle at thermocline depth at the front represents the current flowing along the front (*see* section 6.7).

Figure 6.6 (**A**) Aerial photograph of a tidal mixing front (the one marked C in Figure 6.6). The stratified water is to the right of the front and is a clearer blue compared to the mixed water to the left of the front. (**B**) Water surface view of front showing a CTD rosette sampling directly under the front interface where phytoplankton, feathers and even some seaweed are accumulating.

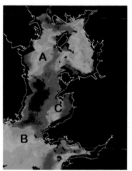

Figure 6.7 Satellite infra-red image of the Irish Sea showing sea surface temperature. Cool surface waters are shown in purple and blue, and warm waters in green, yellow and red. Sharp gradients in sea surface temperature marked at **A**, **B** and **C** are associated with tidal mixing fronts.

$$au^3 = hQ_N \ \text{ or } \ \frac{h}{u^3} = \frac{a}{Q_N}$$

Q_N doesn't change too much from place to place in a given shelf sea area, and so effectively fronts will occur along lines where h/u^3 = constant. Tidal mixing fronts will form at places in the sea where the water depth divided by the cube of the current speed is equal to a certain value. If u is taken as the maximum surface tidal current speed at spring tides, the value of the constant appears to be about 70. So if the water depth is 70 m, we would expect a tidal mixing front to occur along a line where the maximum current is about 1 m s^{-1}. Places where the current is weaker than this would be stratified, and places where the currents are stronger would be vertically mixed.

In fact, as their name suggests, most of the energy at tidal mixing fronts is provided by the tide, and so we can neglect the second term in this equation in comparison with the first. If the stirring is not provided at a sufficient rate, i.e. $au^3 < hQ_N$, complete vertical mixing cannot be maintained and the water column responds by becoming stratified (the rate of increase of potential energy immediately becomes less because only the surface mixed layer is expanding now). Exactly at a tidal mixing front, therefore,

In the Irish Sea, where tidal mixing fronts were first studied, the front runs from the southern tip of the Isle of Man towards the east coast of Ireland (the western Irish Sea front). This separates mixed water to the east (where the tidal currents are strong) from stratified water to the west (where the currents are weak). The other main tidal mixing fronts around the UK are one stretching across the southern entrance to the Irish Sea (the Celtic Sea front), one off the west coast of Scotland (the Islay front), and a long one stretching across the North Sea (the Flamborough Head front). In addition, tidal mixing fronts have been discovered in many shelf sea regions of the world, including George's Bank in Canada, Cook Strait in New Zealand and the Patagonian Shelf in South America.

6.7 Dynamics of tidal mixing fronts

It would be difficult to make the structure shown in Figure 6.5 in a laboratory tank. The warm surface water will tend to spread out towards the left (and the cold bottom water will also tend to spread out along the bottom, also towards the left in the figure). In the sea, this is prevented from happening by the Coriolis effect (Chapter 4). The warm surface water initially spreads out, but then turns to the right (in the northern hemisphere) under the influence of the Earth's rotation. The same thing happens to the cold bottom

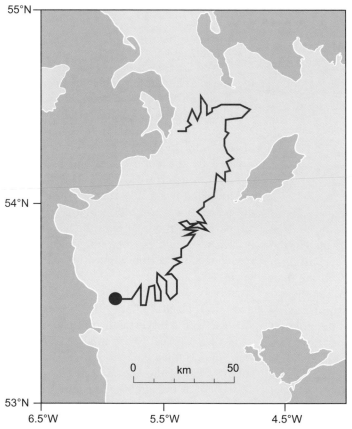

Figure 6.8 Paths of satellite tracked buoys in the Irish Sea. The buoys flow along the front in this part of the Irish Sea, travelling on an anticlockwise circuit known as the gyre.

water. This effect creates currents flowing along the front.

In fact these flows also create a surface slope, and the flow then adjusts so that there is a single current flowing along the front at the depth of the thermocline, at speeds of a few tens of centimetres per second. This flow travels with the stratified water on its left looking along the direction of flow. Because this flow is weak compared to the tidal streams at the front, it is difficult to detect with conventional current meters. It *can*, however, be detected with Lagrangian techniques using satellite-tracked drifters fitted with an underwater sail (or drogue) at thermocline depth. In the western Irish Sea front, such drifters flow along with the stratified water on their left and can perform complete circuits of the stratified water in the western Irish Sea (Figure 6.8). For this reason, this part of the Irish Sea is sometimes called the *gyre*.

We have seen that tidal mixing fronts form along lines where the value of the water depth divided by the tidal current cubed (h/u^3) is constant. The strength of the tidal streams, u, changes with the springs/neaps cycle, and as this happens, the front tries to adjust its position so it remains at the critical value of h/u^3. This movement is called the springs/ neaps adjustment of fronts. At spring tides, some of the stratified water adjacent to the front becomes mixed and the front advances into the stratified water. At neap tides, some of the mixed water adjacent to the front becomes stratified and the front moves back again.

This movement can be observed from space. It is not large – just a few kilometres – but as the front moves back and forth in this way, water is transferred from the high-nutrient mixed water into the low-nutrient stratified water. It is thought that this process may account for some of the high biological productivity observed at these tidal mixing fronts.

7 Light in the oceans

Sunlight is important to life in the sea, as it is on land. Algae suspended in seawater and on the seabed use sunlight to photosynthesise and grow (*see* Chapter 10). In this chapter we examine the quantity and quality of sunlight that arrives at the sea surface and describe how it penetrates into the ocean.

Figure 7.1 The penetration of sunlight into the ocean is critical for life.

7.1 Sunlight

Sunlight is part of the *solar spectrum* (Figure 7.2). A plot of the energy in solar radiation against wavelength peaks at a wavelength of about 500 nm. Visible light occupies the part of the spectrum around the peak – from 400 nm (blue light) to 700 nm (red light). This is also the part of the solar spectrum used in photosynthesis. For this reason, visible light is also called **Photosynthetically Active Radiation** (PAR). It is no coincidence that the most energetic part of solar radiation is called 'visible light' – our eyes have adapted to be sensitive to the part of the solar spectrum that has the most energy. However, it is a coincidence that the ocean is most transparent to visible light. Figure 7.1 also shows the percentage of solar energy that is transmitted without loss through 1 m of clearest seawater. It can be seen that the peak in solar energy and the peak in water clarity are nearly coincident. It is only visible light (and part of the ultraviolet) that is transmitted through seawater to any significant depth. This is a very fortunate coincidence for life in the ocean.

Sunlight can be measured as the amount of energy falling on unit surface area in unit time – a quantity called **irradiance**. The units of energy are Joules, and so we can measure irradiance in $J\,m^{-2}\,s^{-1}$ or $Watts\,m^{-2}$, since 1 Watt (the unit of power) equals 1 Joule per second. The irradiance, at all wavelengths, arriving at the top of the atmosphere has been measured as (about) $1400\,W\,m^{-2}$, a figure called the *solar constant*. At ground level, the irradiance can approach this value at the Equator at noon. At other latitudes, the maximum irradiance at midday is generally much less. For example, Figure 7.3 shows that the total solar energy arriving at a weather station on Snowdon, a mountain in North Wales, on a clear day in September peaks at a value of about $600\,W\,m^{-2}$ at noon.

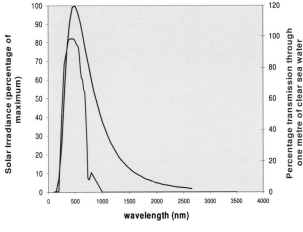

Figure 7.2 The solar spectrum. The blue curve shows the energy in sunlight plotted against wavelength, and the brown curve the transmission through seawater as a function of wavelength.

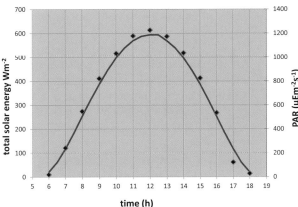

Figure 7.3 Solar irradiance at the summit of Snowdon in North Wales on a sunny day in September. Hourly values are shown as both total solar energy and as a photon count. The continuous curve is a theoretical prediction for the same day based on the solar constant and the changing elevation of the sun through the day.

In *biological oceanography*, it is more common to express the solar irradiance in terms of a photon count. Photons are tiny bundles of visible light energy. A common unit for counting photons in the sea is the µEinstein (or µE), which is 6×10^{17} (or one µmole) of photons. As an approximate guide, 100 Joules of total solar energy contains about 200 µE of photons. We can use this rule of thumb to convert between these physical and biological measures of solar energy. Figure 7.2 shows

the light energy expressed in both these units, using the 2:1 conversion ratio. Photons are counted per unit area, per unit of time and so the commonly used terms are µmol photons $m^{-2}s^{-1}$, or $µE\ m^{-2}s^{-1}$

Solar irradiance varies through the year with the seasons. Figure 7.4 shows the seasonal variation of daily mean irradiance measured at Dunstaffnage in Scotland, from which it can be seen that, at this location, the mean irradiance varies by a factor of about 8 between

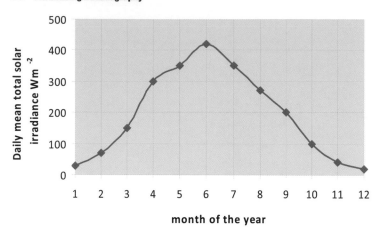

Figure 7.4 Variation of daily mean solar irradiance through the year at Dunstaffnage in Scotland.

winter and summer. This is mostly due to the much longer days at this latitude in summer. At lower latitudes, the shape of the seasonal curve is flatter, and at higher latitudes it becomes more peaked, with daily mean irradiance falling to zero in winter inside the Arctic and Antarctic circles.

7.2 Penetration of sunlight into the sea
The decrease of solar energy with depth in the ocean can be measured with underwater light meters, the most common form of which is called an *irradiance meter* (Figure 7.5). Seawater is in general rather opaque to sunlight, compared to the atmosphere. In cloudless conditions, sunlight travels through the Earth's atmosphere to ground level virtually undiminished. Even in the clearest ocean waters, amounts of sunlight utilisable by photosynthetic organisms rarely penetrate to depths greater than 100 m, and in coastal waters this figure can be less than 10 m.

Figure 7.6 shows an example of measurements of PAR irradiance at different depths at a station in the Clyde Sea in Scotland. The decrease of irradiance with depth, initially

rapid and then slower, is characteristic of an exponential curve in which the rate of change of irradiance at any depth is proportional to the irradiance at that depth. The relationship between the irradiance E at depth z and irradiance at the surface E_0 can be expressed mathematically by the Lambert-Beer Law: $E = E_0 e^{-kz}$

The parameter k, which has units of m^{-1}, is called the **diffuse attenuation coefficient**. The higher the value of k, the more opaque is the water. For a measured set of irradiance values at different depths, such as that shown in Figure 7.6, the diffuse attenuation coefficient can be calculated by fitting the Lambert-Beer law to the data. This is most easily done by plotting the natural logarithm of the irradiance against the depth, which produces a straight-line relationship with a slope equal to $-k$. For the data in Figure 7.6 it can be shown in this way that $k = 0.14\,m^{-1}$. This value is typical of the PAR (or white light) diffuse attenuation coefficient in shelf seas. In clear ocean water, k is much smaller and in turbid water it is greater. For comparison, the attenuation coefficient of visible light in a clear atmosphere is about $0.00001\,m^{-1}$.

Figure 7.5 (**A**) small irradiance sensor designed to measure visible light (or PAR) at different depths under water. (**B**) A sensor attached to the CTD rosette to measure the downwelling light at the same time as the temperature and salinity profiles are being made. Another type of light meter (shaped like a light bulb) can be seen on the apparatus in Figure 1.3a. The difference between the two is that this one measures the light falling on it from above, whereas the 'light bulb' shaped one measures light coming from all around it.

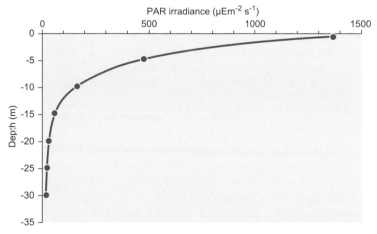

Figure 7.6 Typical decrease in irradiance with depth in coastal waters.

The attenuation coefficient in the sea varies with wavelength. Clear ocean water absorbs red light principally and blue light least. As a result there is very little red light at depth in clear ocean water. Red objects under water appear black because there is no

Physical basis for the exponential decay of light in the sea

The Lambert-Beer Law is consistent with a photon having a set probability of traversing a given path length without being absorbed by the medium through which it is travelling. For example, supposing that the chance of a photon being absorbed in travelling 1 metre vertically down into the sea is 0.5. Imagine we start with 100 photons travelling down through the sea surface. Then, at a depth of 1 m, 50 photons will remain, the other 50 photons having been absorbed in the top metre of the sea. At a depth of 2 m there will be just 25 photons left, at 3 m 13 (since half a photon is impossible) and so on. This decrease in the number of photons, from 100 to 50 to 25 and then 13, which is at first rapid and then slower, is an exponential decay.

red light to reflect off them. Phytoplankton in ocean waters absorb blue light as well as red, and so move the minimum attenuation towards the green. This effect becomes even more apparent in shelf seas where dissolved organic material and suspended mineral particles (or mud) are also effective absorbers of blue light.

The physical process that removes photons as they travel down into the ocean is *absorption*. In this process, the photon hits a particle or a molecule and its light energy is converted into another form of energy. For example, the photon's energy may be used to warm the water or it might be used to drive photosynthesis in a phytoplankton cell. There is another process which affects light travelling down into the sea, and that is *scattering*. Scattering will change a vertical beam of light entering the sea into diffuse light, in which the photons are travelling in all directions, mostly still downwards but not in parallel lines. In diffuse light the average distance a photon travels in reaching a given depth in the sea is greater than that for a beam of light. For this reason, scattering increases the diffuse attenuation coefficient. A small proportion (typically a few per cent) of the light is scattered at more than 90° and travels back up to the ocean surface. It is this back-scattered light, when it enters our eyes, that gives the water its colour (*see* section 7.5).

The depth to which 1% of the incident light is able to penetrate is called the **photic zone**. We can use the **Lambert-Beer Law** to show that the depth of the photic zone in this case is equal to $4.6/k$. As we might expect, as the attenuation coefficient increases, the thickness of the photic zone decreases. For example, an attenuation coefficient of $0.14\,m^{-1}$ is associated with a photic zone extending down to 33 m. An attenuation coefficient of $0.3\,m^{-1}$ would produce a photic zone with a thickness of just 15 m.

7.3 Other ways to measure water transparency

A surprisingly good estimate of the transparency of seawater can be obtained by lowering a white disc into the water and noting the depth at which it just disappears (purists will tell you to go beyond this depth and then note the depth of reappearance as you slowly pull the disc back up). The rope used for lowering can be marked with tape at convenient intervals to help you determine the depth of the disc below the surface. Such

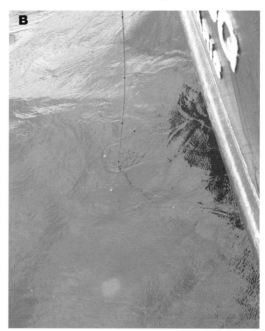

Figure 7.7 (**A**) The Sechhi disc is a simple way of measuring the transparency of seawater. It is lowered into the sea, and the depth at which it disappears (the Secchi depth) is noted (**B**).

a piece of apparatus is called a **Secchi disc** (*see* Figure 7.7) and the depth at which the disc disappears (or reappears) is called the Secchi depth.

As you might expect, there is a relationship between the Secchi depth and the diffuse attenuation coefficient (*see* Figure 7.8). The relationship is an inverse one, since Secchi depth increases as k decreases and the water becomes more transparent. The relationship between Secchi depth and diffuse attenuation coefficient shown in Figure 7.7 can be expressed as:

$$k = 1.4/Zsd$$

where Zsd is the Secchi depth in metres. Other studies give slightly different values for the numerical constant in this equation (it has been suggested that the value of this figure depends on the relative importance of light scattering and absorption in seawater). If we take 1.4 as a working figure and remember that the depth of the photic zone is $4.6/k$ it follows that the depth of the photic zone is just over three times the Secchi depth. The greatest Secchi depth to be published is 80 m! It was measured in the Weddell Sea, Antarctica in October 1986 (i.e. at end of winter when no phytoplankton were in the water). On this occasion the photic depth was around 240 m. Red seaweeds have been found growing attached to rocks at 268 m in waters off the Bahamas, which indicates that the photic zone has extended down that far.

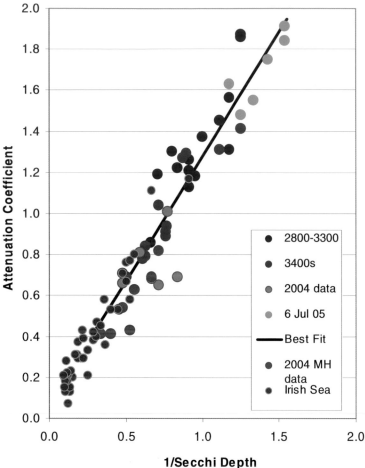

Figure 7.8 The relationship between the attenuation coefficient for white light and the Secchi depth is an inverse one. This diagram shows that there is a good relationship between attenuation and 1/Secchi (both expressed in m^{-1}) depth over a wide range of water clarity values in European waters.

However, these are extreme maxima, and it is generally considered that even in the clearest waters the photic zone does not extend much deeper than 200 m.

A useful instrument for studying the variation of water transparency with depth is the beam transmissometer. This instrument has a light source from which light travels in a parallel beam through the water to a light detector some distance (typically a few tens of centimetres) away. Particles in the water scatter light out of the beam and so reduce the signal reaching the detector. The ratio of the measured signal to that measured in perfectly clear water is called the beam transmittance. Profiles of beam transmittance can be used to study how sediments are suspended from the sea floor. Beam transmissometers can also be placed on moorings to provide information on changes in suspended load with the ebb and flow of tidal currents – useful information for studies of sediment transport.

7.4 The colour of the sea

The colour of the sea, viewed from above, varies from the beautiful deep blue of clear ocean water, through the emerald green of shelf seas to the muddy brown colour of shallow turbid estuaries and coastal water. When viewing the true colour of the sea, it is important not to be confused by the reflection of the sky by the sea surface. This is often what you see when you stand on a cliff and look out to sea, viewing the sea surface at a low angle. The best way to cut out the surface reflection and to view the true colour of the sea is to place one end of a tube just below the water surface and to look down the other end of the tube (Figure 7.9).

The blue colour of clear ocean water is caused by a fundamentally different mechanism to that of the sky. In the atmosphere, light is scattered primarily by air molecules, which are much smaller than the wavelength of light. In this case, a scattering process explored by the English physicist Lord Rayleigh applies.

Blue photons are scattered more readily than those of other colours, and this is what gives the sky its blue colour. In the ocean, scattering of light is caused primarily by suspended particles and inhomogeneities in the water, and the scattering process is more or less wavelength-independent. However, absorption of light in the sea does depend on wavelength. As photons travel into the sea, are backscattered and travel up again to the surface, the more strongly absorbed colours are filtered out, leaving just the most weakly absorbed photons to give the sea its colour. Pure water absorbs red light most strongly and blue light least. This is why clear ocean water has such a strong blue colour – the red and green colours have been removed from the light that is backscattered to the surface (Figure 7.9A).

If phytoplankton (Chapter 8) are abundant, blue light is absorbed as well as red (the chlorophyll pigment in phytoplankton and plants is a good absorber of blue and red light, which is why many plants are green). Phytoplankton living near the sea surface filter out blue light, and water continues to filter out red light, and so the spectrum of the upwelling irradiance at the sea surface shifts towards the green (Figure 7.9B). This is the basis of satellite remote sensing of phytoplankton biomass in the ocean. Satellites can measure irradiance of different colours leaving the ocean surface. The ratio of green to blue irradiance increases as the chlorophyll concentration in surface waters increases. By carefully calibrating the 'blue/green ratio' against measurements of chlorophyll in water samples, it is possible to convert satellite observations of ocean colour into maps of the phytoplankton distribution in the surface waters of the globe (Figure 7.10).

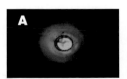
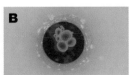
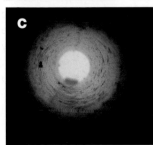

Figure 7.9 Photo of sea water taken down tubes in (**A**) clear blue ocean water; (**B**) phytoplankton rich green water and (**C**) muddy brown coastal water.

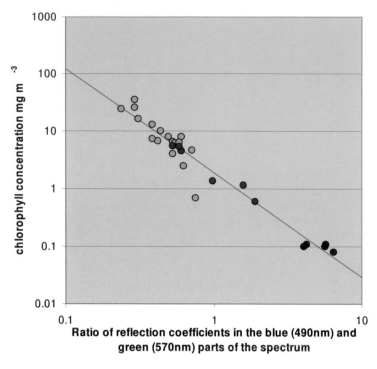

Figure 7.10 Ocean colour measured by satellite can be converted into a chlorophyll concentration, an indicator of algal biomass. This graph shows how chlorophyll in clear Scottish waters and in the Atlantic makes the water greener.

Figure 7.10 shows how, as chlorophyll increases in these waters, the ratio of blue to green reflection coefficients decreases, i.e. the water becomes less blue and more green. This has proved a very useful result for mapping chlorophyll in the world's oceans (see chapter 10 and, in particular, figure 10.5). Unfortunately, the simple relationship shown in figure 7.10 breaks down in shelf seas and estuaries (the most biologically productive parts of the ocean) because here, materials other than chlorophyll can colour the water green, or even brown.

Shelf seas and estuaries contain materials, apart from phytoplankton, that absorb blue light most strongly. These materials include suspended minerals (or mud) and dissolved organic matter (DOM). The DOM is produced by the breakdown of living matter when it dies and decays, and much of the input of DOM into coastal waters is from rivers. These carry high loads of DOM produced originally from agricultural, forest and peat lands and the decay of plant material. A fraction of the DOM has a yellow to brown colour, and is referred to as coloured or chromophoric DOM (CDOM – see figure 7.11). In very turbid water, the absorption of blue and green light by the suspended and absorbed matter can predominate over the absorption of red light by the water itself. Photons that make it back to the sea surface after being scattered in the body of the water are now mostly green and red, with just a smattering of blue and the water appears a muddy brown colour (Figure 7.8C).

Figure 7.11 This sample of estuarine water contains a high concentration of CDOM producing a characteristic brown colouration.

The colour of water will depend, amongst other things, on the concentrations of plankton (mostly expressed as chlorophyll *a* concentration), CDOM and suspended mud in the water. It is also dependent on the optical properties of the water itself. The absorption coefficient of a material is defined as the proportion of light absorbed in travelling a short distance through the material, divided by that distance. It has units of m^{-1}. The absorption coefficients of pure water, and water containing chlorophyll, CDOM and suspended mud particles are shown in Figure 7.12. Since the absorption coefficients of the non-water materials increase with their concentration, it is possible to use the absorption coefficients of water samples, at specific wavelengths, for estimating the concentrations of CDOM and other constituents in the water. This is often less time-consuming and less expensive than doing the chemical analyses directly.

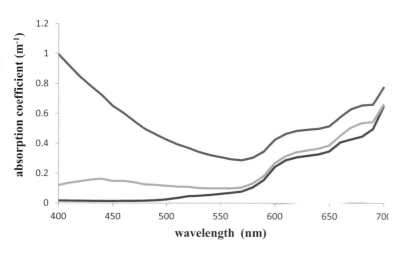

Figure 7.12 Absorption coefficient of pure water (blue); water containing chlorophyll *a* at levels typical of the spring bloom in coastal waters (green) and water containing coastal water levels of CDOM and suspended mud as well as chlorophyll *a* (red).

8 Biology of the oceans

For many the attraction of marine science is the variety of beautiful, and at times simply bizarre organisms that live in the oceans, although thankfully not of the size and ferocity depicted in Figure 8.1. Of course, over and above this the oceans, and especially shelf seas, provide considerable amounts of food for humans to eat.

The whales, at the apex of a complex food web, enjoy a certain charisma among marine animals. A whale can easily weigh around 100 tonnes. This compares with the bacteria at the

Figure 8.1 The creatures of the sea are many and varied, but thankfully not as fearsome as those depicted in this section of the Carta Marina by Olaus Magnus, which was first printed in 1539.

other side of the web, which can weigh around 0.1 picogram (0.1×10^{-12} g) – a difference of 21 orders of magnitude. To help put this into perspective, the total biomass of bacteria in the surface waters of the Southern Ocean is estimated at around 3.2×10^7 tonnes, whereas the biomass of whales in the same waters is only around 8×10^6 tonnes.

The current estimate of total number of species of **eukaryotic** organisms in the oceans (plants, animals, algae, protozoans) is around 230,000. It is thought that in the region of 140,000 of these are known to scientists. In contrast there are estimated to be up to 1,000,000,000 species of bacteria (**prokaryotes**) in the oceans, of which only a minute fraction have been described. So it is very clear that we still have very much to discover about the biota of the oceans.

8.1 Zones of life in the oceans

At the onset of a general description of life in the oceans, it is useful to place the organisms into groups. The most obvious separation of groups is between those that live attached to or within the sediments or benthos (**benthic**), and those that live in the water (**pelagic**). Looking at the former, some benthic habitats are always covered by the water, even at low tide (**subtidal**) and others are periodically uncovered as the tide goes out (**intertidal**). It is important to remember that many of the species living in soft sandy or muddy benthic habitats are not visible, since they burrow into the sediments. In this short introduction we will not really touch on benthic organisms.

The upper 200 m of the pelagic is called the **epipelagic**, which corresponds to the depth of the continental shelf. Underneath this, **mesopelagic** extends from 200 m down to 2000 m, the **bathypelagic** from 2000 to 4000 m, the **abyssopelagic** from 4000 to 6000 m. The deepest zone is the **hadalpelagic**, which extends from 6000 m to the bottom of the deepest trenches around 11,000 m (Figure 8.2).

There is life in every part of the oceans, and sophisticated deep sea missions have brought back video footage of creatures living in the deep ocean trenches, although most of the life at such great depths is microbial (Chapter 12). However, the majority of the marine biota is concentrated in the surface waters, and as a general rule the total mass of biota per unit volume of seawater decreases with depth. This is because most of the food to sustain the food web is produced in the photic zone. Just as on land, the fundament of the food web are those organisms that can photosynthesise, utilising sunlight, carbon dioxide and water to create new organic matter for growth (Chapter 9). On land this is done by the plants, whereas in the ocean this is accomplished by the algae and some bacteria. As described in Chapter 7, the photic zone only extends down to a maximum of between 250 and 300 m in the very clearest of waters, and mostly at a depth of 100 m or less.

8.2 Life in the pelagic zone

The organisms living in the pelagic are crudely divided into the **plankton** and **nekton**. The latter are those species that are strong enough swimmers to be able to move against a water current (e.g. large crustaceans, fish, squid and whales). The plankton are the group of organisms that, although they may be able to swim, cannot swim against an ocean current, and so effectively drift with the water mass (from the Greek, *plantos* that can be roughly translated as 'wanderer'). The organisms that

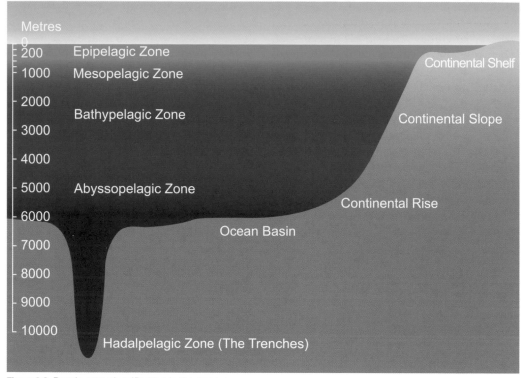

Figure 8.2 Depth zone classification in the oceans.

make up the plankton are generally classified by their size, as shown in Table 8.1. The
other main classification of the plankton is the
organisms that can photosynthesise (mostly
unicellular algae, some flagellates and some
bacteria) called the **phytoplankton**, and the

Table 8.1: Size classes of plankton. It is important to remember that there is considerable overlap between these rough classifications. Organisms less than 200 µm in size can only be seen with good microscopes.

Name	Size range	Examples of typical organisms in each size class
Femtoplankton	0.02 to 0.2 µm	Viruses & small bacteria
Picoplankton	0.2 to 2 µm	Bacteria, flagellates
Nanoplankton	2 to 20 µm	Algae, flagellates, protozoans
Microplankton	20 to 200 µm	Algae, flagellates, protozoans, crustacean juveniles, larvae
Mesoplankton	0.2 to 20 mm	A few algae, crustaceans, small jelly fish, larvae
Macroplankton	2 to 20 cm	Large crustaceans, jelly fish, larvae
Megaplankton	> 20 cm	Jelly fish

non-photosynthesising animals and protozoans referred to as the **zooplankton**.

But even at these most crude of classifications there are exceptions to the rule, since many benthic organisms that live in their adult stage on or in the benthos actually have larval (singular = larva) stages that live in the pelagic. These include fish, molluscs (e.g. mussels), crustaceans (e.g. barnacles or crabs), sponges, worms, echinoderms (e.g. starfish or sea urchins). The release of the larval stages (or **meroplankton**), or gametes (e.g. by seaweeds and corals) into the open waters is a key way that adult populations can ensure that their offspring are dispersed as far as possible, since many of them cannot move very far, or, as is quite often the case, not at all. Organisms that spend all of their life in the plankton are called **haloplankton**.

8.3 Femto- and picoplankton

If you take 1 ml of seawater from most oceans and pass it through a filter with holes of a diameter 0.02 μm (0.02×10^{-6} m), there will be approximately 1 million bacteria cells retained on the filter surface. There will also be around 10 to 100 times more viruses on the filter as well. Viruses are not living organisms, since they cannot grow or replicate themselves without infecting a living cell. However, they are included in this biology of the ocean section, since viral infections can have profound consequences for the rest of the biology. The study of viruses in marine systems is fairly young (two to three decades), but they have been shown to infect most other organisms in the ocean, and during viral epidemics they can result in the very striking death of swarms of crustaceans, algal blooms, seaweeds, fish, whales, etc. Again, to show how important

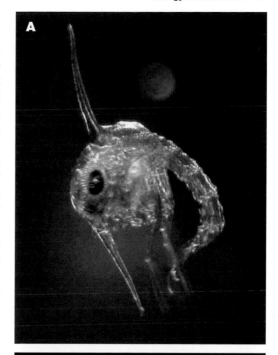

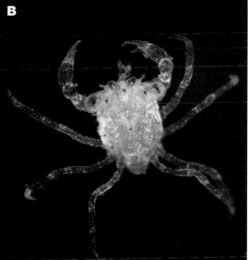

Figure 8.3 Two larval stages of a crab (**A**) Zoea and (**B**) a megalopa.

Figure 8.4 High power microscope image of bacteria stained with fluorescent dye. N.B: this sample has been concentrated, and in normal seawater there would only be one or two cells in this field of view. The big bright spot at the front is a ciliate.

the viruses are – assuming the volume of the oceans is approximately 1.3×10^{21} litres, then there are roughly 4×10^{30} viruses in the oceans. These contain about 200 million tonnes of carbon, which would be the equivalent of the carbon contained in 75 million blue whales (which are about 10% carbon).

Much more is known about the bacteria and **archaea** and their role in marine ecosystems. Both bacteria and archaea are prokaryotes, which means that they do not have a membrane-enclosed nucleus or other cell organelles (cf. eukaryotes, which have complex nuclear membranes and membrane-bound mitochondria and chloroplasts – including the plants, fungi, animals and algae). Many of the bacteria in aquatic systems grow by breaking down organic matter dissolved in the seawater (dissolved organic matter, or DOM, introduced in Chapter 7). The DOM pool in the ocean comes from land via rivers and estuaries (allochthonous DOM), but also from dead and decaying marine organisms, and as faecal matter and secretions of material (e.g. slime) produced by many marine organisms (autochthonous DOM). The significance of bacterial breakdown of DOM will be discussed in chapters 8 and 9. Although bacteria and archaea are ubiquitous in marine waters and sediments, the archaea are noteworthy in that they are frequently found in extreme conditions of high temperature (e.g. hydrothermal vents), hypersaline waters and anoxic sediments.

Some bacteria are able to photosynthesise and utilise sunlight in a similar way to algae and plants (chapters 8 & 9). These include the **cyanobacteria** (formerly called blue-green algae), and there are many species that fall into these small size classes. They are thought to be some of the most numerous photosynthetic organisms on the planet, and are present in most waters, although remarkably, they have only been studied in detail since the 1980s. There are also species of photosynthetic eukaryotes in the picoplankton, although these are even less studied than the cyanobacteria.

8.4 Nano- and microplankton
At the small end of this size range the species are dominated by both photosynthetic and non-photosynthetic **flagellated** organisms. The

latter get their energy by feeding on picoplankton-sized organisms, and they generate feeding currents by moving their **flagella** to waft food items towards them. They are voracious grazers and have been estimated to be able to process 100,000 times their own body volume of water per hour. This is essential, since although there are millions of picoplankton per millilitre of water, living organisms in fact fill only a small fraction of this volume, and the encounter rate between a flagellate and food particle is very low.

There are a host of other grazers ranging from 20 to 200 µm, such as **armoured dinoflagellates** and **ciliates**. The former can even digest prey larger than themselves by exuding a sinister structure called a **pseudopod** that envelopes the prey item and encases it. Enzymes are then released to break down the

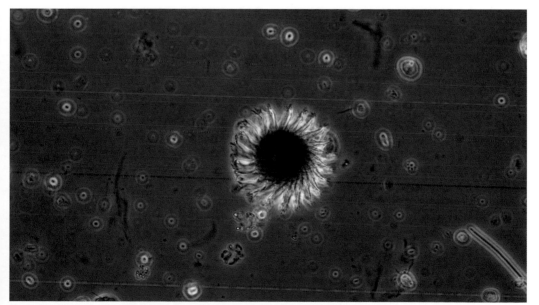

Figure 8.5 High power microscope image of a ciliate. The feather-like cilia create a vortex to concentrate food particles (bacteria and phytoplankton), which are then ingested.

prey, and once this is complete the pseudopod is retracted with the digested remains. Ciliates use another effective technique, by beating numerous hair-like structures (**cilia**) to filter particles out of the water. **Foraminifera**, also unicellular organisms, are another example of grazers that are able to exude pseudopodia (in this case called *reticulopodia*) into the surrounding water to catch their prey. These can extend up to 25 times more than the size of the individual producing them, thereby forming a huge net to catch prey. Foraminifera have shells (**tests**) made from calcium carbonate. In some regions of the world's oceans there are dense accumulations of dead foraminiferal tests (foraminiferal oozes), well preserved in the sediments, that are used by palaeo-oceanographers to describe climate and ocean conditions over geological time.

The most obvious members of the nano- and micro-phytoplankton are the unicellular photosynthetic algae, well represented by a diverse group of species called *diatoms*. The characteristic of all diatoms is that they produce cell walls, or frustules, made of a hard silicate, effectively glass, which are not only very beautiful to look at, but are also very strong. This wall has holes in it, often laid down in very beautiful patterns, which allows water, gases and nutrients to be exchanged. Many diatom species also have highly ornate frustules, with spines, spikes, hooks, and other protrusions. These adaptations are thought to slow down sinking rates, but also to deter grazers from attempting to eat them: the spines of diatoms have even been known to clog fish gills and pierce delicate membranes in gill tissues.

Figure 8.6 Electron microscope image of a dinoflagellate.

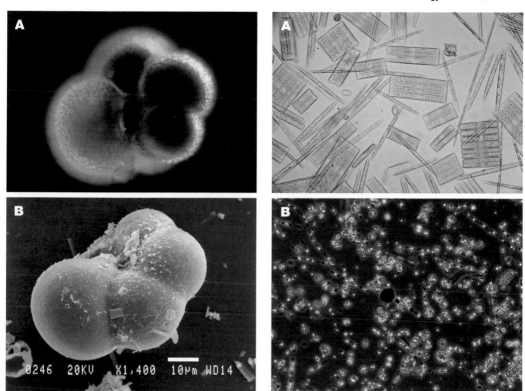

Figure 8.8 (A) Diatoms close up, and (B) during spring bloom.

Figure 8.7 (A) A living pelagic foraminifer, *Neogloboquadrina pachyderma* and (B) a scanning electron microscope picture of the same species preserved in a marine sediemnt. (C) Foraminifer, *Orbulina universa* with extended pseudopodia to catch food.

Diatoms can form dense blooms in coastal waters, and are an important food source for protozoan, and especially zooplankton grazers. These blooms can be so dense and extensive that they are easily seen by colour sensors on satellites (*see* Chapter 14). However, once formed the frustules dissolve very slowly, and some ocean floor sediments are described as diatomaceous or siliceous oozes – massive accumulations of diatom frustules that have sunk from the surface waters. Like the foraminiferal remains described above, these diatom records are prime resources for palaeo-oceanographers for reconstructing past ocean and climate events. Diatomaceous earth can also be mined, and was a key component of the first dynamite invented by Alfred Nobel in 1867.

Diatom blooms are not the only micro-phytoplankton that can be seen from space. The coccolithophores are a group of photosynthetic algal species that are equally beautiful as the diatoms when looked at under the microscope. Instead of being encased in silicate, these single-celled organisms are covered in intricately patterned plates of calcium carbonate (coccoliths). They can also form extensive blooms where the ocean surface turns a milky white. One of the well-known species, *Emiliana huxleyi*, forms blooms in the north Atlantic covering an area of the ocean equivalent to the size of Great Britain. Due to the highly reflective nature of the white coccoliths, these blooms are easily visible from space, and so can be tracked using satellite imagery. The coccoliths sink and are incorporated into sediments, where they can accumulate, locking up huge amounts calcium carbonate. The impressive chalk 'white cliffs of Dover' are largely made up of calcium carbonate from coccolithophores.

8.5 Meso- and macroplankton

There are very few phytoplankton in the mesoplankton, and none in the macroplankton, and so these size classes are dominated by zooplankton species and both halo-and meroplankton larvae. One of the biggest groups of species in this size range is a group of crustaceans (i.e. having an exoskeleton), the copepods, of which there are thought to be over 2000 species. These are hugely important since they are a primary food source for

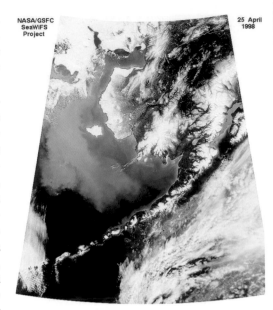

NASA/GSFC
SeaWiFS
Project

25 April
1998

Figure 8.9 Coccolithophore bloom in the Barents Strait that is so extensive that it is visible to satellites.

many fish species and even some whales. Copepods largely feed on the algae and other organisms in the smaller size classes by filtering them out of the water (herbivores or omnivores). However, there are some species that are carnivores, hunting other copepods and having heavily armoured mouth-parts with which to tear apart their prey. Although they cannot swim against currents, they are still good swimmers, and when avoiding predators can swim up to several hundred body lengths per second! Copepods have male and female adults and complex life histories. In some species, after mating the female carries the fertilised eggs in specialised egg sacs, whereas in other species the fertilised eggs are released into the water. When the eggs hatch they release larvae called nauplii. These undergo five further growth phases before reaching a

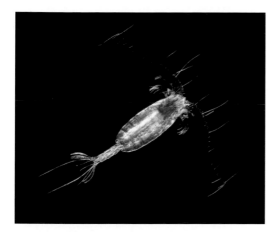

Figure 8.10 Adult copepod.

squid, penguins, many seal species and many baleen whale species.

Some of the most beautiful organisms in this group are the ctenophores (comb jellies) that have rows or fused plates of iridescent cilia which they use to swim with. They are gelatinous multicellular organisms and have specialised guts (therefore different from the ciliates described above, which are unicellular with no specialised guts). Many species have long tentacles that can be extended. These contain sticky substances with which to catch their prey, which includes copepods and other

miniature adult-shaped stage (called a cope-podite stage). As they grow larger the copepodites undergo five more developmental stages before reaching adult maturity.

Towards the larger end of the macroplankton, the crustaceans are also represented by the Euphausiids or krill. These are a diverse group of species that range in size from 1 cm to around 10 cm, the most well-known species of krill being *Euphausia superba*, the Antarctic krill, which can form swarms akin to locust swarms. These swarms can be so dense that they have been recorded as turning the ocean a blood-red colour. Krill are present in such huge stocks in the Southern Ocean that they are estimated to exceed 1.5 billion tonnes (cf. the total mass of all the people on the Earth is approximately 0.5 billion tonnes). Large, dense krill swarms contain up to 30,000 animals per cubic metre, and since each individual can eat up to 25% per cent of its body weight per day, a krill swarm can very effectively clear a water column of any food particles. In turn, krill are the primary food for

Figure 8.11 The Antarctic krill, *Euphausia superba*. In 2009 a megaswarm of krill was discovered by scientists working in Wilhelmina Bay on the western Antarctic Peninsula. The swarm extended from the surface to 400 m deep and was estimated to contain 2 million tonnes of krill. In turn this attracted 500 humpback whales to feed on the swarm.

similar sized organisms. Salps are another frequently encountered gelatinous group in the plankton. They are found in the plankton as either individual transparent tubular bodies, or in long chains of individuals. They are very efficient at filtering particles from the water, especially phytoplankton cells.

It would be normal to think of molluscs (snails and bivalves) as primarily benthic organisms. However, all benthic molluscs have larval stages that live as meroplankton. Moreover, there are several groups of molluscs that are haloplankton, and even some that

have heavy shells. A good example of the latter are the pteropods (winged foot), whose feet (the part of a snail that is normally used to move along a solid object) are flattened into a set of paired 'wings' with which they swim. Some pteropods produce mucous floats to help them keep suspended, whereas other species have lost the heavy shells altogether, although they still use mucous nets to help keep them suspended and for catching prey.

Some of the most fearsome predators of the plankton world are the chaetognaths (arrow worms) which can be between 1 and

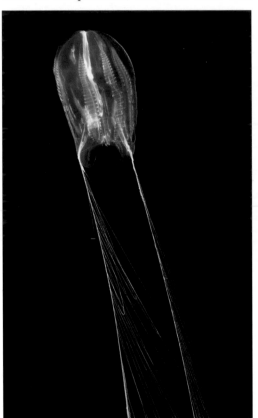

Figure 8.12 Ctenophore or comb jelly.

Figure 8.13 A pteropod or 'flying snail'.

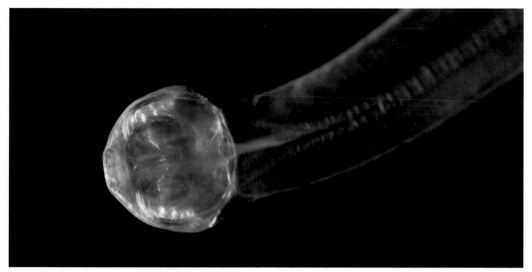

Figure 8.14 The mouth of a chaetognath, a voracious predator of zooplankton and small larvae.

10 cm long. They are very commonly found in samples of plankton. They are transparent, streamlined in shape, and are effective swimmers that can dart forward with great speed to catch their prey (often copepods and larvae). Some species of arrow worm also use powerful neural toxins to kill their prey before ingesting them.

It is pertinent to point out that many species of zooplankton are good enough swimmers that they take part in a diurnal vertical migration (DVM); many populations of zooplankton are observed to move up from depths between 100 and 400 m at dusk and down again at dawn in a highly synchronised way. The most popular explanation for this is that by spending periods of strong light during the day at depth means that they avoid predators like fish, birds and squid that use vision to seek prey. Then by moving up to the surface at night they are able to feed. However, it is not necessarily so simple, and many ecologists have spent considerable effort in trying to unravel the secrets of this phenomenon. For instance, one theory is that the zooplankton migrate to deeper waters in the day to avoid damaging ultraviolet radiation. Clearly the primary trigger to induce the behaviour is light, but temperature is also thought to play a role, and seasonal differences are observed in some water bodies where DVM takes place in the summer but not in winter.

Some species of copepods are also known to migrate to deeper waters to overwinter when food is scarce in the surface waters. They enter a physiological state called **diapause**, where they become physically inactive, don't feed, reduce their metabolism to a minimum and sink to deep waters where they are neutrally buoyant (where their density is the same as that of the surrounding seawater). Depending on the species, the copepods can overwinter in this state at depths between 500 and 3500 m.

8.7 Megaplankton

The most noticeable megaplankton are the jellyfish (belonging to the group Cnidarians, which also includes sea anemones, corals and the comb jellies mentioned above). Although jellyfish swim effectively with rhythmic contractions of their bell, they are unable to swim against ocean currents, and so drift where the water masses take them. This is surprising, since several species can reach huge sizes, for instance the Arctic 'lion's mane' jellyfish (*Cyanea arctica*) which has tentacles up to 40 m long.

Jellyfish are mostly carnivores feeding on zooplankton, larvae and small fish. To capture their prey many species trail tentacles laden with specialist cells called nematocysts. When touched, these cells can release a hollow, harpoon-like structure that anchors the prey (zooplankton and larvae) and injects a powerful toxin to immobilise it.

Many of the smaller jellyfish species, more typical of macroplankton, have a plankton medusa phase, which alternates with a polyp stage that grows attached to the benthos or structures in the sea. The polyps produce and release larval medusae that eventually develop into the more visible jellyfish.

Some jellyfish species, such as the siphonophores, are colonial without any polyp stage. They are remarkable in that the individuals within the colony adopt different functions for propulsion, feeding, or reproduction. In the Portuguese Man O'War (*Physalia physalis*) one part of the colony develops into a large gas-filled float (up to > 30 cm high) which keeps the colony suspended at the water surface. It is hard to see one of these structures and to realise that it is not a single organism. These also have tentacles laden with nematocysts, which in the largest of specimens can trail below the float for several tens of metres.

The potency of the toxins released by the nematocysts is well known to any swimmer or diver who has had the misfortune to touch a tentacle. The encounter results in what can be an intensely painful experience. However, the toxins produced by the sea wasp or box jellyfish (*Chironex fleckeri*) are so potent that they have caused several human deaths. For this reason, beaches are often closed if they are spotted in the water.

8.8 Nekton and others

The marine organisms of the nekton are better known and include fish, reptiles, squid and mammals, including the toothed and baleen

Figure 8.15 A compass jellyfish caught in the Namibian Benguela, where the biomass of this species now exceeds that of commercially important pelagic finfish. NB the deck in the background is awash with them.

whales, sea otters, manatees and dugongs. There is a whole group of other marine animals that are clearly able to swim against a horizontal current, and so are nekton, but which can also have periods out of the water. These include polar bears, seals, sea lions, saltwater crocodiles and turtles. Clearly, fish and squid are only nekton when adults, since their egg and/or larval stages take place among the plankton.

It is impossible to do justice to such a variety of animals in just a few short paragraphs.

Figure 8.16 Many of the larger animals of our oceans spend time outside the water, but are all part of the nekton (**A**) Polar bear, (**B**) Leopard seal.

However, by being able to swim so effectively many of these organisms, such as the baleen whales and many species of fish, are able to roam large distances on tremendous migrations through many ocean realms. Naturally, by being able to swim effectively, some of the animals are able to live in deeper waters and migrate to surface waters just to feed.

Finally in this section, it would be wrong not to introduce a key group of organisms, other than humans, that feed on the animals of the nekton. Although not part of the nekton, there are the myriad of bird species that exist by feeding on fish and zooplankton mostly in surface waters of the oceans (0 to 50 m deep). However, some, such as the Emperor Penguins (*Aptenodytes forsteri*) are such exceptional swimmers that they can dive to depths of 500 m in search of fish, squid and krill. The rich diversity of bird species and the huge biomass of birds that the oceans support mean that they are an integral part of the ocean ecosystem.

8.8 Living and sinking in a viscous medium

The experience of movement in water is very different from that in air, since water is more *viscous* than air. The viscosity of the medium (the resistance of a fluid to flow under the influence of an applied external force) is dependent on the density of the water, and hence temperature and salinity. A term called the **kinematic viscosity** is a measure of the drag on objects moving through the fluid, and for that object inertia tries to keep the object moving, whereas viscosity tries to stop it. Basically the greater the viscosity of a medium, the more slowly an object will move through it; for instance, olive oil has a much greater viscosity than water, and a stone will sink more slowly in the olive oil compared to the water.

Figure 8.17 Birds are an important part of the open ocean ecosystem, e.g. (**A**) Giant petrels that seem to appear out of nowhere when some debris is caught up on the water surface. (**B**) Emperor penguins, (**C**) Gannets, and (**D**) Puffins.

For the movement of any particle in a specific medium we talk about the **Reynolds Number** (RN) which is derived from:

RN = (particle velocity × particle size)/kinematic viscosity

Very crudely, the smaller the organism the smaller the RN: a large whale swimming in seawater at $10\,\mathrm{m\,s^{-1}}$ has an RN of 300,000,000; a large fish swimming at the same speed an RN of 30,000,000; a copepod swimming at $0.2\,\mathrm{m\,s^{-1}}$ an RN of 300; and a bacteria swimming at $0.01\,\mathrm{ms^{-1}}$ an RN of 0.00001.

The consequence of this is that for small organisms the fluid is effectively highly viscous, and a moving object has little momentum: e.g. when a bacteria stops swimming its momentum will keep it coasting for less than 10 microseconds (around 1 nm), whereas a large fish or a whale will continue to glide a considerable distance through the water, even though it has stopped active swimming. This lack of momentum on the micro-scale means that when you look at micro-organisms through a microscope they appear to be moving through oil or thick syrup.

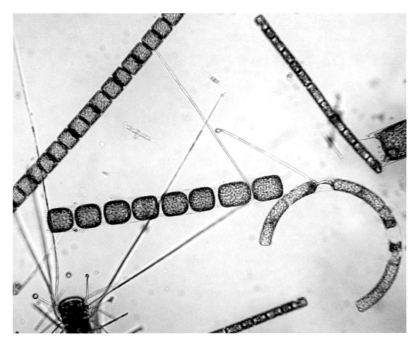

Figure 8.18 As shown here, the spines and shape of diatoms will greatly influence their rate of sinking through the water.

Naturally, these effects translate into how fast different-sized organisms sink in the water column, and this is of immense consequence, for instance, to photosynthetic organisms wanting to stay in the photic layer. **Stokes' law** describing falling spheres can be used to calculate the terminal velocity of an organism at low Reynolds numbers (effectively for particles/organisms up to 500 μm in diameter). The rate of fall of a sphere (of radius r) is described by the equation:

$$V = \frac{2}{9} \frac{(\rho'-\rho)}{\rho v} gr^2$$

where V = velocity of particle, g = acceleration due to gravity, ρ' = density of particle, ρ = density of water, v = kinematic viscosity of water. Therefore a particle's rate of sinking increases in proportion to the square of its radius. Observations of sinking rates of phytoplankton cells actually show that their sinking speed increases with cell radius raised to a power lying between 1 and 2. The reason is that in many cases the density (ρ) of the cell decreases with cell size, and this lowers the power to which the size is raised in this sinking speed equation.

9   Chemistry of the oceans

Normally in any course or book introducing marine science/oceanography the physics is described first, then the chemistry and lastly the biology. However, in this short introduction the focus will be on the chemistry that directly influences the biology, and so it was important to introduce some of the key groups of organisms in the previous chapter. Marine organisms are greatly influenced by the chemical composition of the medium surrounding them and, in turn, can modify it. Some can tolerate a wide range of chemical conditions, whereas others only have a narrow tolerance to changes in the chemical composition of the surrounding environment. Microbial activity drives many of the chemical processes routinely monitored by marine chemists, and this realisation has led to increasingly closer union between biologists and chemists in their study of the oceans, the dynamics of biology and the elemental flow between the living and non-living reservoirs – so much so that it is popular to talk about the study of ocean biogeochemistry (although the term is at least 80 years old). The 'bio-' and '-chemistry' parts of this word are clear, but the '-geo-' bit in the middle is there because the processes being studied are important for the cycling of elements over geological timescales. They also produce chemical signatures that are important for interpreting the geological record in rocks and ocean sediments, or even in long-lived organisms, such as corals.

9.1 Salty water

As said in Chapter 1, ocean waters are salty and the concentration of salts in the water is measured as salinity. The average oceanic salinity is about 35, although in coastal waters it can be closer to 0, and in some regions (e.g. the Red Sea) give up to 40. However, there are parts of the ocean where the dissolved salts become very concentrated, such as in brines trapped in sea ice with salinity exceeding 100, which are produced when seawater freezes, and in hypersaline bodies of water (salinity >250) which have been discovered in deep-water basins in the eastern Mediterranean Sea and are thought to be tens of thousands of years old, and are also characterised by having no oxygen (anoxic). Despite the large range in salinity in oceanic waters, the ratios between the major constituents of seawater (Table 9.1) remain constant, and they are said to behave conservatively. Most of the naturally occurring

Table 9.1: Major constituents of seawater at a salinity of 35.

Ion	% by weight
Chloride (Cl^-)	55.03
Sodium (Na^+)	30.59
Sulphate (SO_4^{2-})	7.68
Magnesium (Mg^{2+})	3.68
Calcium (Ca^{2+})	1.18
Potassium (K^+)	1.11
Bicarbonate (HCO_3^-)	0.41
Bromide (Br^-)	0.19
Borate ($BH_2O_3^-$)	0.08
Strontium (Sr^{2+})	0.04
Everything else	<0.01

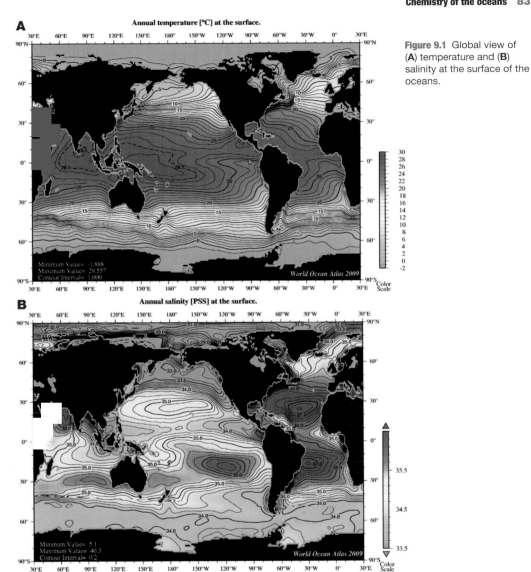

A

Annual temperature [°C] at the surface.

Minimum Value= -1.888
Maximum Value= 29.557
Contour Interval= 1.000

World Ocean Atlas 2009

B

Annual salinity [PSS] at the surface.

Minimum Value= 5.1
Maximum Value= 40.3
Contour Interval= 0.2

World Ocean Atlas 2009

Figure 9.1 Global view of (**A**) temperature and (**B**) salinity at the surface of the oceans.

elements (around 90) have been measured in seawater, but the largest fraction of the ions in seawater is made up of fewer than ten elements (Table 9.1).

The conservative relationship between the constituents of seawater and salinity provide a useful tool for looking at the behaviour of material entering coastal water from rivers through estuaries. Progressing from a river mouth towards the sea normally results in the salinity gradually increasing from 0 to the salinity of the coastal water. If the concentration of a particular ion is greater in river water than in seawater and no other processes occur in the intervening estuary than simple physical mixing of the riverine water and the seawater masses, the concentration of the ion will decrease with increasing salinity in the estuary (dilution line, Figure 9.2A). Conversely, if the concentration of a particular ion is greater in seawater than in the river water, there will be an increasing concentration with increasing salinity in the estuary when physical mixing of the two water masses is the only process there (Figure 9.2B).

However, if the values in the estuary plot below the dilution line (physical mixing), this is an indication that some additional process occurs in the estuary that removes the ion from the water. Similarly, if the data is above the dilution line, it indicates that some additional process has added more of the ion to the estuarine water.

9.2 What does 'dissolved' mean?

Talking about ions dissolved in water is rather straightforward, but on the broader scale, seawater is a complex salty chemical *soup*, also comprising the products from the death and decay, excretion and defecation of all the organisms in the ocean. The ions of elements in seawater and the compounds they can form are termed 'inorganic', whereas the biology and its products are organic (shells, etc. are inorganic). In terrestrial systems most dead animal and plant material, and waste products are incorporated into the soils and are broken down there. In contrast, in the oceans many of the equivalent processes are taking place in the water column. So there are carbohydrates, lipids, proteins, amino acids, and complex compounds of many molecular sizes in seawater.

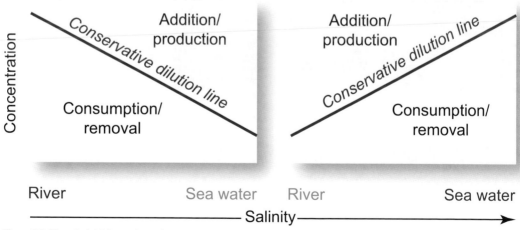

Figure 9.2 The straight lines show the conservative mixing plot of the concentration of a substance vs. salinity. (A) Concentration of a particular ion or substance is greater in river water than in seawater (B) Concentration of a particular ion is greater in seawater than in the river water.

In normal oceanographic practice it is common to use filters with pore sizes between 0.2 and 0.45 μm to separate between dissolved and particulate fractions (that passing through the filter is dissolved, and that retained on the filter is particulate). Commonly used terms for the particulate and dissolved organic phases are dissolved and particulate organic matter (DOM and POM). However, the dissolved fraction through such size cut-offs must contain some bacteria and viruses, since as we saw in the last chapter, the smallest bacteria and viruses are in the 0.02 to 0.2 μm size range. In practice routinely filtering seawater through filters with such tiny pore sizes is impractical, since they become clogged very quickly.

Some phytoplankton and bacteria secrete complex sugar molecules (polysaccharides) for protection against changes in external conditions. This material is called extracellular polymeric substances (EPS) and can be present in seawater in considerable amounts. The EPS, visible only when stained with special dyes, are gels which change the viscosity of the water, especially important for the smaller size classes of plankton. They are of undefined shape and size, and span the continuum between the dissolved and particulate phases (they can range from micrometres to centimetres). These gels also provide surfaces on which bacteria grow, so that they can become hotspots of bacterial activity, since many more bacteria are concentrated into a smaller volume than is normal in the open water.

When looking at the carbon in the waters of the oceans (not the sediments) the dissolved organic carbon (DOC) pool is by far the largest pool of organic carbon. At first this is rather surprising, since the particulate organic carbon pool (POC) includes all organic particles and organisms from viruses through

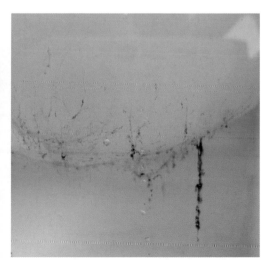

Figure 9.3 EPS gels produced by diatoms are stained here in blue.

Figure 9.4 Foam washed up onto a beach is not a nasty chemical spill, but rather whisked up dissolved organic matter (DOM).

to blue whales. But there are about 700 giga-tonnes (giga=x10^9) of carbon as DOC and only 30 gigatonnes of POC. Of the POC only around 3 gigatonnes consist of living organisms.

Sometimes there is so much DOM in the water that when it is sufficiently agitated, as in during a storm, the DOM is whisked up to form a foam that washes up onto beaches, frequently to the alarm of the public (Figure 9.4). There is, of course, nothing to worry about – it is just the organic compound soup being churned up and the proteins and carbohydrates frothing up, just like when whisking milk to make a cappuccino coffee.

9.3 Gases in seawater

As temperature and salinity increase, the solubility of gases in seawater decreases. As pressure increases (with increasing depth) gases become more soluble. A good illustration of these principles happens when a bottle of carbonated (fizzy) drink is opened. In the factory the liquid is injected with carbon dioxide under high pressure, and when the bottle was sealed in the factory the pressure inside the bottle was very high. Upon opening the bottle top the pressure is released, which causes a lot of gas to come out of solution (in bubbles) since the pressure is lowered. If the bottle were warmed up, more bubbles of gas would be released, and throw in some salt, even more so. So following these principles, polar oceans should have higher concentrations of dissolved gases than those at the tropics, deeper waters higher concentrations, and more saline waters lower gas concentrations than those in fresh waters. As we will see, this is not always the case.

The four main gases in air are nitrogen (N$_2$ – 78.1% by volume), oxygen (O$_2$ – 20.9%),

argon (Ar – 0.9%) and carbon dioxide (CO$_2$ – 0.03%), although here we will only deal with O$_2$ and CO$_2$. Typically when in contact with the air the gas content of a liquid changes so that it reaches an equilibrium with the concentration, or rather, partial pressure (the pressure the gas would have if it occupied the volume alone), of the gas in the air.

However, in the case of CO$_2$ it is not so straightforward, since in the ocean the CO$_2$ reacts with water to form carbonic acid:

$$CO_2 + H_2O \Leftrightarrow H_2CO_3$$

Carbonic acid in turn dissociates rapidly to form the bicarbonate ion:

$$H_2CO_3 \Leftrightarrow H^+ + HCO_3^-$$

Furthermore, the bicarbonate ion dissociates further to form the carbonate ion:

$$HCO_3^- \Leftrightarrow H^+ + CO_3^{2-}$$

So in seawater dissolved inorganic carbon occurs as dissolved CO$_2$ gas, carbonic acid (H$_2$CO$_3$), and bicarbonate (HCO$_3^-$), and carbonate (more CO$_3^{2-}$) ions. The proportions of these in seawater are in an equilibrium (see equation above) that is primarily governed by the pH (but also salinity and temperature) of the water:

$$CO_2 + H_2O \Leftrightarrow H_2CO_3 \Leftrightarrow H^+ + HCO_3^- \Leftrightarrow H^+ + CO_3^{2-}$$

In seawater of a salinity of 35 and a 'typical' pH of around 8, around 90% of the inorganic carbon occurs as HCO$_3^-$, and only about 0.5% in the form of CO$_2$ gas (*see* Figure 9.5). In more acidic waters (lowering of pH) the shift of this equilibrium is to the left (more CO$_2$ gas) and in more alkaline waters (increase of pH) the shift is towards the right (more CO$_3^{2-}$). So if you drop some concentrated acid into some

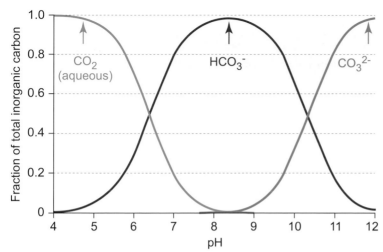

Figure 9.5 The form of CO_2 in the water is strongly dependent on pH.

seawater, shifting the pH to below 2, the whole equilibrium (9.4) shifts to the left.

However, in normal ranges of oceanic pH if CO_2 gas is removed from the water the disturbed equilibrium will shift to the left in 9.4, resulting in the other ions changing concentration until more CO_2 is produced and equilibrium is re-established. NB the total amount of carbon will still have gone down because of the lost CO_2.

The case for O_2 is not as complex as that of CO_2. The oxygen simply dissolves in seawater, although often its concentration in the surface of the ocean is in excess of that expected from the concentration in the seawater being in equilibrium with the air. This is because waves on the ocean surface cause bubbles of gas to form and be carried down into the water column, and because of the increased pressure the gases dissolve in the water, causing them to be present in supersaturated concentrations. The other reason for the supersaturated levels of O_2 is a result of photosynthesis by phytoplankton that occurs in the photic layer. Photosynthesis uses CO_2 and produces O_2 (Chapter 10), and when there is a lot of photosynthesis (only in the light) this further increases the O_2 inputs into the photic layer.

Respiration, which is carried out, day and night, by all marine organisms from bacteria to whales, utilises O_2 and produces CO_2 (again see Chapter 10 for more details). In particular, when bacteria break down decaying organic matter (DOM and POM) as it falls through the ocean, considerable O_2 is consumed and CO_2 is produced.

All of these processes combine to produce commonly found O_2 profiles with depth in open ocean waters: high concentrations (often supersaturated) in the upper waters (top 100 m or so) and then decreasing to an oxygen minimum between 200 and 1000 m, due to O_2 being utilised by respiring organisms (there is also a corresponding maximum in CO_2 concentration). However, the concentrations do not continue to fall, as would be expected since bacteria and animals continue to respire right down to the sea floor. This is

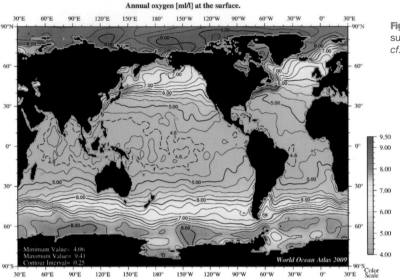

Figure 9.6 Oxygen in the surface of world's oceans; *cf.* Figure 9.1.

because much of the deep water of the oceans was formed in the polar oceans, and is therefore very cold and contains high concentrations of oxygen (*see* chapter 1). The low oxygen zones in the water profile are called the oxygen minimum zone (OMZ). The OMZ regions of the world are thought to be a lower boundary that restricts the depth to which pelagic fish with high oxygen demands (e.g. tuna) are able to dive. At low oxygen concentrations, typical of OMZs, the fish may become physiologically stressed and so have to avoid these areas of the water column. Outside of OMZs, the maximum diving depths of electronically tagged blue marlin (*Makaira nigricans*) are greater than in waters with a pronounced OMZ.

Sometimes, especially if there is a lot of organic matter being broken down by bacteria, such as after a dense algal bloom, the oxygen mimina can be quite low (suboxic). However, in shallower coastal waters, especially those

with little exchange of water from other regions, if there is strong stratification, the respiration below the photic zone may be so great that it actually utilises all the available oxygen, and the system becomes anoxic. When this happens it is a disaster, since it leads to the death of all those organisms requiring oxygen to respire – from algae and bacteria through to fish. NB marine mammals are not killed because they have to come to the water surface to breathe.

There is yet another twist to the complexity of the carbonate dynamics in seawater. In the last chapter, coccolithophores were introduced as being important bloom-forming phytoplankton. They are covered in coccoliths made of calcium carbonate, and when calcium carbonate structures form, one of the products is CO_2. So although the coccolithophores are taking up CO_2 through photosynthesis, they are producing it when they respire and as they produce their distinctive outer coverings:

$$Ca^{2+} + 2HCO_3^- \Rightarrow CaCO_3 + CO_2 + H_2O$$

9.4 Nutrients in seawater

Among the 'everything else' in the major constituents of seawater in Table 9.1 are the main nutrients that phytoplankton require for growth (*see* Chapter 10). This alone indicates that they are not there in unlimited quantities, and in fact in many parts of the oceans they can be present in such low amounts that phytoplankton cannot grow. Conversely, especially in coastal regions with a lot of run-off from agricultural land, they can be present in such quantities that the growth of phytoplankton they support is a problem (Eutrophication, *see* Chapter 13).

Although many nutrients are important for phytoplankton growth, the most commonly limiting major ones are nitrogen and phosphorus, exactly as on land, and the reason why farmers add nitrogen and phosphorus fertilisers to their arable crops. However, the land–ocean comparison is a weak one because even in the waters containing the highest concentrations of nutrients, their concentrations are a tiny fraction of the concentrations in most unfertilised soils.

Nitrogen is present in seawater as dissolved N_2 gas, ammonium (NH_4^+), nitrite (NO_2^-), nitrate (NO_3^-), and in a range of organic molecules. N_2 gas cannot be used except by a few cyanobacteria that are able to **fix** the nitrogen so that it can be incorporated into amino acids and eventually protein. The main nitrogen source used by phytoplankton is NO_3^- and to a lesser extent NH_4^+. Phosphorus occurs in several inorganic forms in seawater, although at pH 8 HPO_4^{2-} accounts for most of the free phosphate ions, and it is the form of inorganic phosphorus that is most bio-available for cell uptake and metabolism.

Clearly since the nitrogen and phosphorus are not sufficiently abundant, and the inputs from rivers are not enough to sustain continued phytoplankton growth over years, there must be another source of these nutrients. This comes from the excretion of urea and NH_4^+ by zooplankton and nekton, and from the breakdown of dead organisms and faecal material by bacteria in the water column, as well as the sediments on the sea floor. The bacteria break down the proteins and amino acids to produce NH_4^+ which is subsequently converted to NO_3^- by a group of bacteria called nitrifying bacteria. This whole process of converting organic matter into inorganic nutrients is referred to as nutrient regeneration or remineralisation. Phosphate is also released back into the water through bacterial activity breaking down organic matter.

In the deep oceans the surface waters have generally low concentrations of NO_3^- and HPO_4^{2-}. Although often associated with the oxygen minima between 500 and 100 m, there are peaks in these nutrients due to the high amounts of bacterial activity going on that have led to the oxygen depletion. Generally waters below surface mixed layers have higher concentrations of nutrients, and it is when mixing events occur, bringing deep waters to the surface, that nutrients are injected into the photic zone. In shallow shelf seas where seasonal thermoclines break down in autumn and winter, it is then that the surface waters become replete with nutrients that are of key importance for the following spring bloom (Chapter 10).

Where waters are low in nutrients they are said to be oligotrophic. Where there are high concentrations of nutrients the waters are said to be eutrophic, and in between the two they are mesotrophic.

9.5 Interconnected oceans

An enormous movement of water, which is called the global ocean conveyor belt or the thermohaline circulation (Chapter 1), interconnects the oceans. As millions of square kilometres of seawater freeze in the Arctic and Southern (Antarctic) Oceans (Chapter 12), cold, highly saline brines are expelled from the growing ice sheets, increasing the density of the water and causing it to sink.

In the conveyor belt circulation, warm surface and intermediate waters (0 to 1000 m) are transported towards the Arctic in the north Atlantic, where they are cooled and sink to form North Atlantic Deep Water, which flows southwards. In the Southern Ocean ice

formation also produces cold high-density water that sinks to form Antarctic Bottom Water. These deep-water masses move into the South Indian and Pacific Oceans, where they rise towards the surface. The return leg of the transport begins with surface waters from the north-eastern Pacific Ocean flowing into the Indian Ocean and then into the Atlantic. This is not a fast process since, if it were possible to tag a molecule of water at the start in the north Atlantic, it would be several thousand years before it got back to the start again.

It is not just the temperature and salinity of the deep water formation in the polar regions that is crucial to the ocean circulation. As said above, the oxygen-rich waters from the poles are essential for ensuring that the deep waters remain well oxygenated despite all the bacterial activity going on to break down organic matter. But also, the longer the water masses are away from the surface of the oceans, the longer time there is for the bacteria to drive

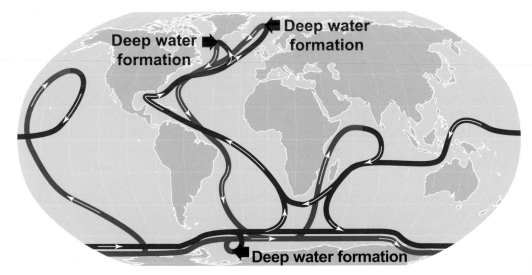

Figure 9.7 Sketch of the global thermohaline circulation showing the regions where deep bottom water is produced.

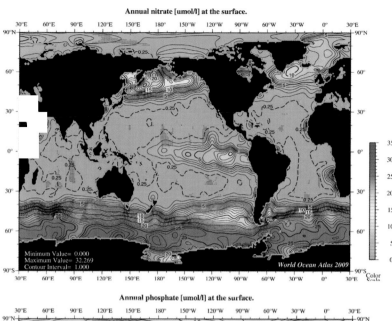

Annual nitrate [umol/l] at the surface.

Minimum Value= 0.000
Maximum Value= 32.269
Contour Interval= 1.000

World Ocean Atlas 2009

Figure 9.8
Global distribution of
nitrate in the surface of the
oceans.

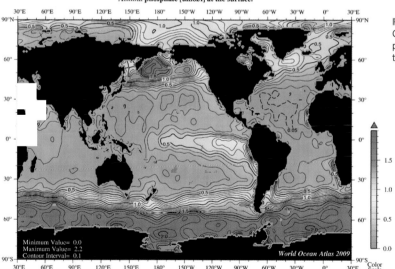

Annual phosphate [umol/l] at the surface.

Minimum Value= 0.0
Maximum Value= 2.2
Contour Interval= 0.1

World Ocean Atlas 2009

Figure 9.9
Global distribution of
phosphate in the surface of
the oceans.

these remineralisation processes. As a consequence of this, the waters coming towards the surface in the north-eastern Pacific have much higher concentrations of NO_3^- and HPO_4^{2-}, (and lower O_2 concentrations) than north Atlantic waters.

10 Primary production in the oceans

As stressed before in previous chapters, the fundamental process underlying all of the marine foodwebs is the production of organic matter through photosynthesis by the phytoplankton and some members of the picoplankton. The amount of new biomass that is made by the photosynthetic organisms (primary production) ultimately determines the size of zooplankton, fish, whale and seabird stocks, as will be discussed in Chapter 11. So understanding the factors controlling primary production in the oceans is of primary importance in our understanding of the management and utilisation of the biological resources of the oceans.

10.1 Photosynthesis and respiration

In the oceans there are very few plant species. Seaweeds, which cover many intertidal shores and grow down to 50 m or so in the subtidal zone, look as though they should be classified as plants. However, they are in fact algae (macroalgae or large algae). Although they photosynthesise, they differ from plants in that they do not flower and have no roots, leaves or highly organised tissues for transporting water and nutrients. The only true plants in the oceans are a relatively small group of the sea grasses. The unicellular algae of the phytoplankton are related to the seaweeds and termed microalgae (or microscopic algae).

These issues aside, the photosynthetic plankton can photosynthesise in the same way that terrestrial plants do, in that they use light as their energy source and carbon dioxide to produce new organic matter, a consequence of which is the production of oxygen. The photosynthetic process can be divided into two: a light reaction and a dark reaction. The light reaction converts light into metabolic energy and reducing power. Specialised light sensitive pigments such as chlorophylls absorb light energy, which is subsequently used in the dark reaction to reduce (fix) CO_2 to organic compounds (generally symbolised as CH_2O). The overall reactions are:

Light reaction: $2H_2O + light \Rightarrow 4[H^+] + metabolic\ energy + O_2$

Dark reaction: $4[H^+] + metabolic\ energy + CO_2 \Rightarrow [CH_2O] + H_2O$

Generally photosynthesis (combining light and dark reactions) is represented by the following equation:

$6CO_2 + 6H_2O + 48photons\ of\ light \Rightarrow 6O_2 + C_6H_{12}O_6$

The most common pigment used by phytoplankton, as in plants, to trap light is chlorophyll a (Chl$a$), which absorbs blue (maximally at 430 nm) and red wavelengths (maximally at 680 nm) of light and reflects green wavelengths (the reason why plants look green). However, there is a large variety of other chlorophylls and other pigments that are used in different species to absorb other wavelengths of the PAR (400 to 700 nm, *see* Chapter 7): Carotenoids

Figure 10.1 (**A**) Seaweeds are macroalgae which are related to (**B**) Phytoplankton. Both are not plants but are algae.

such as β-carotene and fucoxanthin, as well as chlorophyll-b, absorb in the green part of the light spectrum (400 to 520 nm), whereas phycoerythrin absorbs in a different range of the green region (490 to 570 nm). Phycocyanins and allophycocyanins absorb light in the green–yellow (550 to 630 nm) and orange–red (650 to 670 nm) parts of the spectrum respectively. These pigments are examples of accessory pigments, and phytoplankton adjust the composition and concentrations of their light harvesting pigments to absorb various components of the PAR spectrum of light as it varies with water depth. So a phytoplankton cell in well illuminated surface water will have a different composition of pigments than that found near the bottom of the photic zone. Additionally, as phytoplankton move from a high light to a region of low light they synthesise more chlorophyll and pigments to be able to maximise the harvesting of photons. Conversely when moving from low light to high-light conditions they tend to decrease the concentration of chlorophyll and other pigments inside the cells. Some of accessory pigments are very effective at screening out ultraviolet (UV) parts of the light spectrum (which can cause immense damage to cells), and often in high-light environments UV screening pigments are very rapidly produced.

The other essential metabolic pathway for life to proceed is respiration. All organisms respire to produce energy, and unlike photosynthesis, which can only take place in the light, respiration takes place continuously; but critically it is dependent on there being oxygen present. Effectively the equation for respiration is the reverse of that for photosynthesis:

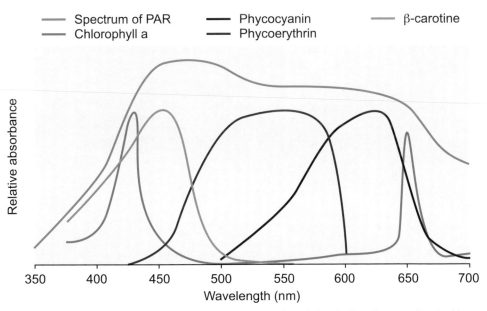

Figure 10.2 Spectrum of PAR with absorption spectra of chlorophyll and other pigments sketched in.

$6O_2 + C_6H_{12}O_6 \Rightarrow 6CO_2 + 6H_2O + energy$

The amount of energy produced is highly dependent on the nature of the starting material: for a given weight of material, the respiration of lipids (fats) produces at least twice as much energy as carbohydrates (sugars), which is why fats are widely used as a storage material in all groups of life.

10.2 Phytoplankton growth

A phytoplankton cell is not just composed of the carbohydrates generated by photosynthesis, but is made up of cell walls, membranes, proteins, enzymes, etc. So as well as photosynthesis taking place, inorganic nutrients such as nitrogen, phosphorus, and sulphur need to be assimilated to build the complexity of life. The total list of elements that are needed is enormous, although some of these are required in minute or trace amounts. As a very crude estimation, a functioning phytoplankton cell comprises around 40% protein, 5% nucleic acids and nucleotides, 40% carbohydrates, and 15% lipids. This allows us to rewrite the simplified photosynthesis equation to take into account the need for nitrogen and phosphorus:

$106CO_2 + 16NO_3^- + HPO_4^{2-} + 122H_2O + 18H^+$
$\Rightarrow C_{106}H_{263}O_{110}N_{16}P + 138O_2$

From the equation above we can derive the ratio of carbon: nitrogen: phosphorus in healthy, actively growing algal cells as 106:16:1. This ratio is the same as the Redfield ratio (after the American oceanographer A.C. Redfield) which describes the elemental composition of the bulk of the particulate organic matter in the oceans. The C:N ratio in the above is 6.6:1, which is a commonly used measure to determine the physiological status

of algae, since when nitrogen is limited, or the algal cells are senescent or dying, this ratio increases considerably.

10.3 Photosynthesis and light

The relationship between photosynthesis and light (irradiance) is described by a photosynthesis/irradiance curve (P/I curve, Figure 10.3). It is important to remember that although photosynthesis is light-dependent, respiration is constant. With increasing light, photosynthesis increases linearly (with a slope of α) until it reaches a particular irradiance where the photosynthetic rate is equal to the respiration rate. This point is called the **compensation irradiance, Ic**. The irradiance where this occurs is species specific, and even within a single species can vary seasonally or on even shorter timescales.

As irradiance increases further, the trend becomes gradually non-linear and a point is reached where further increases in irradiance do not result in increases in the photosynthetic rate. In other words, the rate of photosynthesis is light saturated, and the maximum rate of photosynthesis (P_{max}) has been reached. The saturation irradiance I_k is the irradiance at the intercept of the two lines shown in figure 10.3. In some organisms, there is a decrease in photosynthetic rates at very high irradiances (photoinhibition) resulting from damage to cell membranes or proteins.

Two other important terms need to be noted: gross photosynthesis is equivalent to the total photosynthesis, and net photosynthesis is equal to gross photosynthesis minus respiration. The distinction between these is pertinent, since it is only really the net photosynthetic energy gains that are available for cell growth and reproduction.

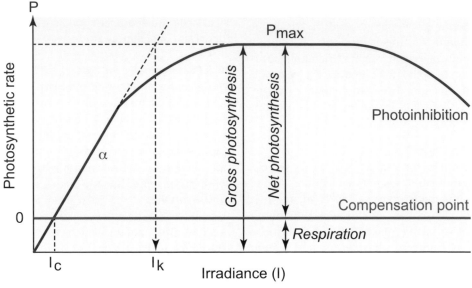

Figure 10.3 Photosynthesis vs. Irradiance curve (P/I curve).

10.4 Photosynthesis and growth

By comparing what we know about how light decreases through the photic zone, and coupling this with the information of the P/I curve, it is possible to describe the trends in photosynthesis in the surface oceans. If a cell could maintain its position at the very surface of the water, naturally this is the maximum light available for optimum photosynthesis to occur. If it is at the bottom of the photic layer (1% of incident light), there may still be enough light for photosynthesis to occur, but potentially not enough to reach or surpass the compensation point. There will be a point in the water column, the compensation depth, at which the gross photosynthetic carbon assimilation by the phytoplankton equals the respiratory carbon losses, or when the net photosynthesis is 0 (equivalent to I_c on the P/I curve).

However, phytoplankton cells are not static in the water, since they are sinking and being mixed either throughout the whole water column or, where water stratification takes place, within surface mixed water layers (Chapter 3). The depth of the mixed layer may be below that of the compensation depth, and so when considering net phytoplankton growth it is more pertinent to relate the daily integrated photosynthetic gains to the integrated respiration losses over the water column (day and night) to the depth of the mixed layer.

The critical depth is the water depth where the integrated daily photosynthetic carbon assimilation is balanced by the integrated daily respiratory carbon losses. As long as sufficient nutrients are present, net phytoplankton growth occurs when the mixed layer depth is shallower than the critical depth. When the mixed layer extends below the critical depth algal growth is limited by light, and there is no net phytoplankton growth.

If the water is so turbid that very little light reaches the bottom of the mixed layer, it can be shown that the mean irradiance experienced by the cell is equal to $I_0/(kh)$, where I_0 is the surface irradiance, h is the depth of the mixed layer and k is the attenuation coefficient. If this mean irradiance is greater than the I_c for the cell, then it can grow in the surface mixed layer.

In the spring, the seasonal thermocline becomes established in the sea in temperate latitudes, trapping phytoplankton in a relatively shallow surface mixed layer (*see* Chapter 6). At the same time, the surface light levels are increasing. As a result of the increasing surface irradiance (I_0) and the decreasing depth of the surface mixed layer (h), the mean irradiance in the surface mixed layer increases sharply in the spring (Figure 10.4). Often, a clearing of the water in the surface layer as suspended sediment sinks out, reducing the attenuation, will also help to increase mean irradiance in the surface layer. The *spring bloom*, when phytoplankton start to grow rapidly, producing a visible greening of the water, begins when the mean irradiance in the surface mixed layer exceeds the compensation irradiance.

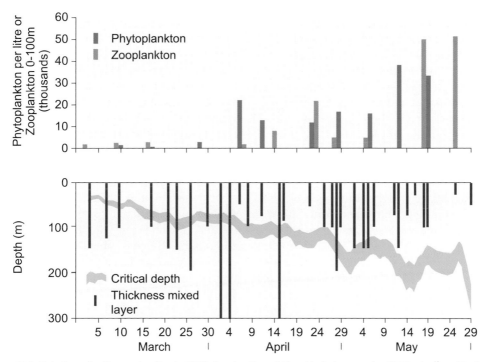

Figure 10.4 Data from the Norwegian Sea in 1949 showing the relationship between mixed layer depth, critical depth, and phytoplankton and zooplankton abundance. Growth of phytoplankton only occurred when the depth of mixing was consistently above the critical depth. (Illustration adapted from Sverdrup's original data).

Example calculation of the timing of the spring bloom

As a quantitative illustration of the occurrence of the spring phytoplankton bloom, suppose that the compensation irradiance for a certain phytoplankton group has a value of $20\,\mu E\,m^2 s^{-1}$. Let the depth of the surface mixed layer be 30 m and the diffuse attenuation coefficient be $0.5\,m^{-1}$. If the surface irradiance is I_0, then the mean irradiance in the surface mixed layer will be $I_0/(0.5*30) = I_0/15$. The spring bloom will start on the day on which this mean irradiance first exceeds the compensation irradiance, that is $I_0/15>20$, from which it follows that $I_0>300\,\mu E\,m^{-2}s^{-1}$. If these conditions were occurring at the latitude of Dunstaffnage, we would expect the spring bloom to start in the middle of April (see figure 7.4).

10.5 Nutrients limiting phytoplankton growth

Phytoplankton growth can only carry on as long as all the nutrients are present in adequate supply, and seasonal algal blooms are normally brought to an end when a nutrient (normally phosphorus or nitrogen) becomes used up. In the scenario described above for temperate shelf seas, nutrients are returned to surface waters in autumn and winter once the thermocline breaks up and the nutrient-depleted surface waters are able to mix with the nutrient-rich deeper layers. If there is enough light still around when this breakdown of the thermocline begins (which can sometimes be the case in autumn) there may be a second short bloom of phytoplankton.

Generally we talk about nitrogen or phosphorus limiting phytoplankton growth; however, it may not be the lack of one of these major nutrients that stops the growth, but rather one of the nutrients needed in only trace amounts. There are some ocean waters where phytoplankton growth is never sufficient to deplete the nitrogen and phosphorus in surface waters. These are known as high-nutrient, low-chlorophyll (HNLC) waters, and are located in the subarctic Pacific, the Southern Ocean, and the equatorial Pacific. What is common for all of these is that they are relatively far away from land. Iron is an essential element for several metabolic processes in phytoplankton cells, but it turns out that these waters have remarkably low concentrations of dissolved iron. The primary source of iron to the surface waters of the oceans is from the land, either directly or via deposition of dust from the atmosphere. Over the past 20 years there have been a number of large-scale experiments in several HNLC regions where dissolved iron has been added to the surface waters, and these have all resulted in spectacular phytoplankton growth. So although only needed in trace amounts, the iron is clearly the limiting nutrient in these systems.

As we have discussed, in stratified waters, algal growth in the upper mixed layer can deplete the nutrients to the point where further growth is not possible. However, a commonly observed phenomenon in such circumstances is a layer at, or just above, the thermocline where phytoplankton growth is still taking place. These layers are called sub-surface chlorophyll maxima, and form when the boundary between the surface-mixed waters (nutrient-depleted) and lower water mass (nutrient-replete) is above the critical depth. Nutrients from waters below the boundary layer diffuse upwards, aided by

Figure 10.5 Chla distribution in surface waters of the oceans. Note the large areas in the subarctic Pacific, the Southern Ocean, and the equatorial Pacific where productivity is low, the so-called HNLC regions.

small-scale turbulence, into the thermocline region. Phytoplankton entrained in this zone are able to use the nutrients since there is still enough light for net growth to take place.

This is an example of how phytoplankton growth is being stimulated by the coming together of two water bodies with different physical and chemical properties. The surface waters are stabilised, warm and well lit, but lack nutrients, and the waters below the thermocline are cold and dark but do have nutrients. Therefore at the front between the water layers the two water bodies complement each other to provide conditions favourable for primary production to take place. Similar complementation effects are seen at other frontal systems in shelf seas, such as tidal fronts or shelf-sea fronts, and fronts associated with river/estuarine plumes.

There are several coastal regions, notably off Peru, Chile, south and north-eastern coasts of Africa, where winds blowing parallel to the coastline result in the transport of nutrient-depleted surface waters away from the land. This is replaced by nutrient-rich water from below (upwelling), the consequence of which is that these regions support the highest primary production, and in turn zooplankton and fish production, on Earth.

Upwelling of nutrient-rich waters also occurs at major ocean frontal systems in the open oceans, such as the equatorial regions in the Pacific Ocean where the trade winds generate two westerly flowing surface currents, the North and South Equatorial Currents. The Coriolis effect causes the currents to deflect northwards in the northern hemisphere and southwards in the southern hemisphere. The divergent flow of these surface waters away from the Equator causes nutrient-rich water to upwell to the surface, supporting higher rates of primary production than that in adjacent waters. **Cyclonic gyres** found in several regions (anticlockwise in the northern

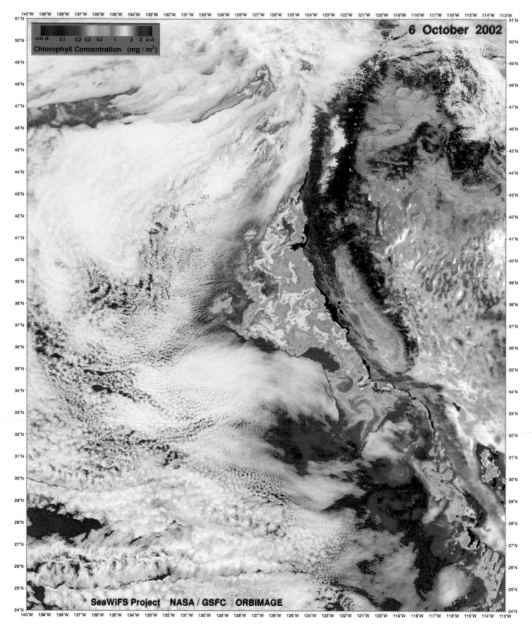

Figure 10.6 Upwelling of nutrient-rich waters off the coast of California causes increased primary production as revealed by this satellite image of surface chlorophyll concentrations.

hemisphere and clockwise in the southern hemisphere) also result in nutrient-rich waters being upwelled from below the thermocline into surface waters as a result of the Coriolis effect producing slopes in the pycnocline, as described in section 4.4. Higher rates of primary production are a consequence of this nutrient exchange.

10.5 Primary production on a global scale

Trying to classify annual primary production of all the oceans is of course difficult, but remote sensing techniques have enabled us to make good estimations of the regional differences (Figure 10.7). It is quite evident that as well as upwelling regions, the coastal waters and epipelagic overlying continental shelves also support the greatest primary productivity.

It would be expected that due to the good light conditions in tropical and sub-tropical oceans, these regions would be highly productive. However, this is not the case, since there is normally a permanent thermal stratification, which means that there is no mixing with deeper waters for long periods of time, and the upper mixed layers are generally depleted in major nutrients. In complete contrast, in the polar oceans there is generally significant exchange of nutrient-rich lower waters with the surface waters, and these regions are characterised by long periods of low light and sea-ice cover which severely limits primary production. However, when the light does come, day lengths are long and sun angles high, so there tend to be short but intense periods of primary production in the Arctic and Southern Oceans, although remembering that the latter is an HNLC region and iron limited.

In conclusion, estimates of the net primary productivity of the global ocean range between 40 and 50 Pg carbon year^{-1} (P = *peta*, and 1 Pg is equivalent to 10^{15} g). This value is remarkably similar to the estimates of total land-based primary production. In other terms, the productivity of algae and photosynthetic bacteria in the top 200 m of the oceans equals that of ALL the vegetation in the forests, grasslands and crops on land.

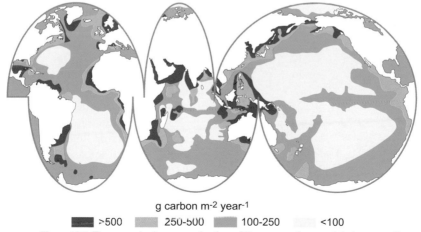

g carbon m-2 year-1

■ >500   ■ 250-500   ■ 100-250   <100

Figure 10.7 The annual primary production of the oceans from a global perspective.

11 Ocean food webs

So far in this introduction, the biology has concentrated mostly on factors leading to producing phytoplankton growth. However, there is clearly much more in the oceans than microscopic unicellular algae and bacteria, and this chapter will consider how the energy trapped by the phytoplankton goes to fuel the various food webs that can ultimately give rise to animals as huge as the whales (Fig. 11.1).

11.1 Growth and food chain efficiency

Ultimately the growth of any organism is a result of the balance between the energy input and the energetic output for life to take place. Some energetic gains, investments, and losses of an organism as it grows include:

Material and energy gains: Photosynthesis (for algae); food intake (for animals);
Material investments: Creating new cells, tissues or skeleton formation, production of energy storage compounds and formation of reproductive material; and
Energy and material losses: Movement, buoyancy, feeding, excretion, water content regulation.

Naturally all this energy comes from the breakdown and respiration of the products from photosynthesis in algae, or the respiration of ingested food in other organisms. The efficiency by which the energy taken in is converted into growth is called the **growth yield**, and is normally somewhere between 10 and 30%. Clearly the amount an organism expends on moving will be a large factor in this. This 'loss of energy' or 'inefficiency' can also be used when considering the **transfer efficiency** of energy between organisms in a food chain. Even the process of feeding in itself can be a huge drain on energy. It has been estimated that Right Whales (*Eubalaena glacialis*) need to consume at least 4500 copepods m^{-3} to balance the energy they expend in feeding.

Many studies have shown that there is a strong positive relationship between the amount of phytoplankton and the total

Figure 11.1 Whales in phytoplankton bloom – start to finish of food chain.

production of fish and squid. However, behind every tonne of fish landed a remarkable amount of phytoplankton production needs to occur: As can be seen from the example in Table 11.1, if the growth efficiency at each organism level is 10%, for every 10 g of the *enormous fish* there would have to be 1 tonne of phytoplankton to support it. Even if the efficiency were as high as 30%, 1 tonne of phytoplankton would still only ultimately produce 2.4 kg of the *enormous fish*. Scaled up to global terms, it has been estimated around 45×10^9 tonnes of phytoplankton are produced globally per year. In contrast, total fish production is probably less than 2×10^8 tonnes per year.

This discussion illustrates why fish species, such as anchovies and sardines, that feed directly on phytoplankton are more productive and abundant than fish such as tuna and sharks that are at the top of a longer food chain, more akin to the example given in Table 11.1. The basic rule is that the shorter the food chain, the more efficient it is. For instance, humpback whales (*Megaptera novaeangliae*), which can weigh over 30 tonnes, feed on krill in the Southern Ocean and are therefore the top of a very efficient food chain: phytoplankton ⇒ krill ⇒ whale. Another efficient food chain producing a large organism at the end is that of the basking shark (*Cetorhinus maximus*) that can grow to over 10 m in length and weigh more than 4 tonnes, by eating only plankton.

Figure 11.2 The energy expended in searching for prey can be quite considerable in the open ocean, very much reducing the growth yield of an organism.

Figure 11.3 Humpback whales feed on zooplankton, and the food chain resulting in a whale is shorter than that of the dolphin shown in Figure 11.3, which eats fish as its main prey.

Steps in food chain	Yield %	Loss %	Example in weight (kg)
1. Phytoplankton			1000
2. Zooplankton	10	90	100
3. Small fish	1	99	10
4. Medium-sized fish	0.1	99.9	1
5. Large fish	0.01	99.99	0.1
6. Enormous fish	0.001	99.999	0.01

Table 11.1 Example of how the yield and loss terms at each level of a food web interact to reduce the production (weight) of the final product. 1000 kg of phytoplankton are needed to 'produce' just 0.01 kg of the enormous fish

As another general rule, pelagic animals will eat organisms whole, and therefore the size of the mouth is key in determining the size of the prey that can be eaten. For this reason predators normally prey on food considerably smaller than themselves, although this is not true in the microbial world (*see* dinoflagellates in Chapter 8) and clearly not true of the basking shark! It has been estimated that fish prey on food that has a body mass >400 times lower than their own. This is why the animals at the top of the marine food chains tend to be bigger. Naturally, this discussion is only pertinent to omnivorous and carnivorous species.

Of course the normal situation is normally far more complicated that the straightforward food chains described here. Instead of chains, quite often there are complex food webs describing the transfer of energy through to the top predators. Figure 11.4 shows how complex things can really be.

Figure 11.5 Shows a more understandable food web for the North Sea herring, which shows that there are several 'pathways' by

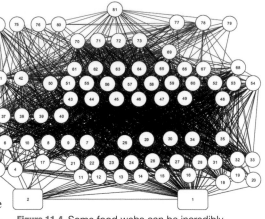

Figure 11.4 Some food webs can be incredibly complex, like this one that includes benthic and pelagic species making up the north-west Atlantic food web. 1 is detritus, 2 phytoplankton, 74 baleen whales, 75 toothed whales, 76 seals, 77 migratory **scombrids**, 78 migratory billfish, 80 birds and 81 humans.

which energy gets grom the phytoplankton to the herring, although the most direct path is the short chain of phytoplankton ⇒ copepods ⇒ herring.

An important item missing from the picture in Figure 11.5 is the part of the food web known as the **microbial loop**. This is the pathway

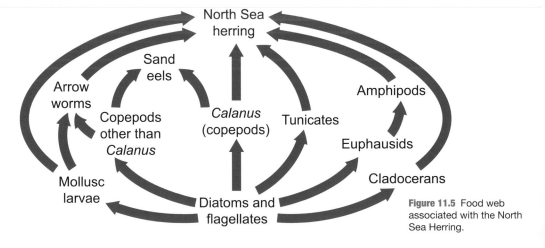

Figure 11.5 Food web associated with the North Sea Herring.

described in Chapter 9, where organic matter (both dissolved and particulate) produced by all levels in the food web is broken down by bacteria, a consequence of which is the regeneration of nutrients for new phytoplankton growth. This was largely overlooked until 30 years ago, but now receives considerable research attention, since ultimately it is this unseen part of the web that makes everything else possible.

11.2 What happens to the phytoplankton

In Chapter 10 the spring phytoplankton bloom was rather left drifting, and the consequences of it not followed up. The phytoplankton of the spring bloom are eaten by zooplankton and small nekton (like the sardines and anchovies). These can only grow once the phytoplankton have bloomed, and so there is normally a peak in zooplankton biomass following the peak in the phytoplankton bloom. Some species of copepods end their diapause (Chapter 8) in early spring before the phytoplankton bloom, migrating to the surface waters where they spawn, thereby ensuring that the young nauplii are in the surface waters to feed on the phytoplankton bloom when it starts. After the zooplankton numbers reach a maximum, predators such as fish larvae move in to feed, and subsequently larger predators feed on the

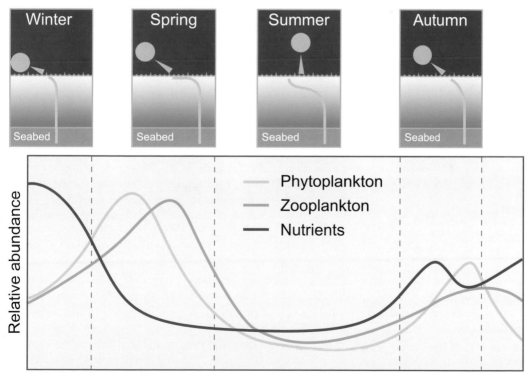

Figure 11.6 Seasonal development of thermocline and corresponding phytoplankton and zooplankton dynamics.

fish larvae, and so on up the food chain (Fig. 11.5). It is not just the magnitude of the phytoplankton bloom that is important, but also the timing of it, since it is thought that blooms that occur early in the season may reduce the amount of fish larvae starvation and so ultimately influence the number of adult fish.

In conclusion to this section it is worth considering the so-called *paradox of the plankton*. It is normally accepted in ecology that when resources are limited (such as nutrients for the phytoplankton) and species are having to compete for them in a homogenous medium, a single species will end up dominating. However, in the plankton, despite the limited resources, there are many species co-existing, an apparent ecological paradox. The concept was introduced by G.E. Hutchinson in 1961 and has been debated long and hard by general ecologists and biological oceanographers ever since. It is beyond the scope of this introduction to join the debate, but clearly the complexity of biological (photosynthesis, EPS, nutrient acquisition, grazing) and non-biological (light, temperature, stratification, etc.) factors that go into determining the nature of planktonic life bring into question the idea of the plankton living in a homogenous medium.

11.3 Fueling the deep sea

All the organisms living below the photic zone are ultimately dependent on food coming down from above, if they cannot swim up to get it themselves. So clearly deep seas below highly productive surface waters have more food available than those beneath less productive waters. But even so, a proportion of the dead organisms from any surface layer never make it to deep waters. As discussed in Chapter 9, this is because they are broken down by bacteria. It is estimated that the flux of organic carbon reaching the sea floor in oligotrophic deep oceans is less than 3% of the surface primary production. Also the quality of the food changes as it sinks, since amino acids and proteins are used up faster than carbohydrates or lipids. Therefore, the material reaching the sea floor is relatively poor in terms of quality, unless it has been transported rapidly through the water with little chance for the bacteria to break it down.

Of course how long a particle takes to sink to deep waters depends on how quickly it sinks through the water column. Some diatom species, for example, have been measured as sinking between 0.1 and 10 m per day, although this naturally depends on the size and shape of the diatom, and it seems that older or dead diatoms sink faster than actively growing ones. So this means that it could take anywhere between 10 and 1000 days for the cells to settle through 100 m of water.

However, if sediment traps are placed in the water to collect the particles falling through the water column, a lot of the material is in the form of faecal pellets produced by zooplankton (Figure 11.7). These pellets are bound by thin membranes and contain the remains of phytoplankton and bacteria, and are effectively packages of organic matter measuring in diameter from a few to several hundred micrometres. The larger the grazer, the larger the faecal pellet, although not all faecal pellets are equally durable; e.g. krill pellets are somewhat fragile and break up rapidly, whereas pellets produced by many of the small copepod species are far more robust and long-lasting. Faecal pellets can sink through the water at fast rates, varying from around 50 to 200 m per day, and even rates of up to 1,500 m per

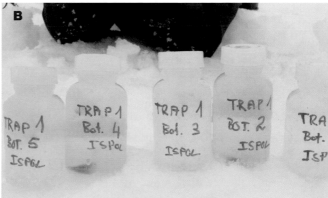

Figure 11.7 (**A**) Small sediment trap being deployed under an Antarctic ice floe. Note the bottles at the bottom of the trap, which were programmed to collect material for four days. (**B**) The bottles after recovery of the trap showing the material that has collected in them.

Figure 11.8 Range of faecal pellets collected in a sediment trap deployed 400 m below the surface of the ocean.

day have been measured. The numbers of faecal pellets *raining down* can be impressive, especially in productive waters where large numbers of copepods have been feeding, reaching values of up to >300,000 pellets m^{-2} day^{-1}. Clearly such pellets greatly speed up the flux of organic matter from the photic zone to the deep sea.

Quite often the contents of sediment traps are in a sort of amorphous gel. This is because much of the particle flux down the water column falls as marine snow. In chapter 9, EPS was introduced; a sticky mucus-like substance produced by many phytoplankton and bacteria. This stuff can form clumps, and marine snow particles are accumulations of EPS often encasing numerous dead or dying phytoplankton cells, bacteria and remains of other plankton. The size of these marine snow agglomerations can range from the micro-metre to centimetre size range, although in extreme cases monster particles with diame-ters up to metres have been seen. It is thought that marine snow is the main vector by which matter is transported to the deep sea.

The sinking material – marine snow, algal cells, bacteria and faecal pellets – that even-tually makes it to the seafloor is collectively called phytodetritus. When there has been a massive fall of material, such as following an algal bloom, this material can form a loose, fluffy layer up to several centimetres deep on top of the sediments. This, of course, is a source of food for benthic suspension and deposit-feeding organisms. However, when the falls of phytodetritus are huge they can actually be a problem, since in general food supply is so poor that a sudden downfall of organic rich material can overwhelm organ-isms more accustomed to periods of austerity.

Figure 11. 9 Aggregate of marine snow approximately 5 mm in diameter. The aggregate based around an abandoned house of a larvacean (free-swimming filter feeder), with many phytoplankton and detrital particles on the filters. Such houses are a major source of marine snow in coastal and oceanic waters.

Figure 11.10 Phytodetrital layer (about 2.5 cm deep) on top of sediment core retrieved from the deep sea.

Also, naturally bacterial activity is high in the rich organic soup like phytodetritus, which can consume considerable amounts of oxygen. If this is high enough it can lead to localised spots of suboxic or even anoxic sediments.

Of course not all particles sinking through the ocean are small, and dead fish and mammals also sink to the ocean floor. It is thought that a whale carcass can provide food for a large array of scavengers and other organisms for periods of decades or more. Of course a decaying whale carcass will be a massive hotspot for microbial activity as well. Naturally it is very difficult to witness a whale fall, but as more researchers are able to study the deep sea, it is clear that, although unpredictable, they are important sources of nutrition to an otherwise food restricted zone.

The transport of carbon as organic matter from the surface of the oceans to the deep sea is referred to as the biological pump, where it may be isolated from the atmosphere for

Figure 11.11 Whale fall images. (**A**) 'Fresh' carcass and (**B**) the remains of a carcass after several years.

hundreds or even thousands of years. Ulti-mately the material reaching the sea floor is incorporated in one form or another into the sediments, although, due to the microbial pro-cesses, only a tiny fraction of material reaching the depths is actually turned permanently into sediments (sequestered). Considering how slow sedimentation rates are, it is remarkable

to contemplate the historical record that they represent, and this must be part of the fasci-nation for marine geologists. However, the biological pump, which is ultimately what has been discussed in the past few chapters, is a way of transferring carbon from the atmos-phere to burial in the sediments for potentially millions of years.

12 Biology at the ocean extremes

One of the theories prevalent in the mid-1800s was that there was no life below around 500 m in the oceans, the so-called **azoic hypothesis**. This was quickly dispelled as soon as people started looking more carefully and systematically sampling deep waters, or just observing the encrustations on deep-sea cables when they were brought up for repair.

We now know that there is no part of the ocean that doesn't support life: organisms are found in the frozen waters and ice of the pack ice; in and on the sediments of the deepest marine trenches; in and around hydrothermal vents. The organisms living in these habitats at the extremes of the temperature range tend to be small, and viable bacteria have been found in sea ice at $<$-20 °C in the Arctic and at $>$100 °C in hydrothermal vent fluids. Naturally these bacteria are highly adapted to these conditions, but although the temperatures seem extreme to us, for the organisms themselves they are not extreme, they are the

Figure 12.1 Intertidal rocky shores exposed to high levels of wave action are among the most 'hostile' habitats for marine organisms

norm. Arguably, organisms living on an inter-tidal shore, or shallow rock pool, are subject to a more extreme existence than those living on the deep ocean floor. Wave action, highly variable air temperatures and extremes of rain or evaporation (changing the surrounding salinity) really do require an organism to be able to adapt rapidly to unpredictable changes in salinity, temperature and light.

12.1 Frozen oceans

One of the most dramatic changes that takes place in the Southern and Arctic Oceans is the transformation of open ocean into frozen wastelands that look more like landscapes than seascapes. This is the pack ice (or sea ice) that covers vast expanses of the polar oceans for much of the year. It is an important platform for the wildlife commonly associated

with these frozen climes: Polar bears and Arctic foxes searching for food roam large dis-tances over the Arctic sea ice during winter. In the Antarctic, penguins and seals use the ice surrounding the continent as a refuge from predators and for migration. In both the Arctic and Antarctic seal species use the ice for a platform on which to give birth and to raise their young.

Underneath a pristine white snow cover, the ice is often a light brown through to rich coffee colour, caused by a thriving **sea ice biota** – a multitude of tiny (mostly microscopic) organisms dominated by diatoms, but also comprising bacteria, fungi, copepods, amphi-pods, turbellarian and nematode worms. They include many of the phyto- and zooplankton of the open ocean, which is in fact where they originate. But unlike their counterparts in

Figure 12.2 Weddell Seals and Adélie penguins hauled out on sea ice.

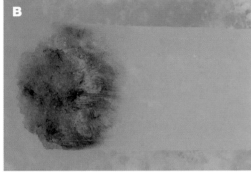

Figure 12.3 (**A**) Brown ice floe and (**B**) bottom of core taken through ice floe.

warmer seas and oceans, these polar planktonic organisms have a unique phase in their seasonal cycle when they are caught up into the semi-solid matrix of the ice.

The transition from a free-ranging life in the open ocean, to one trapped inside an icy layer on the ocean's surface begins when freezing winds rapidly cool the top few metres of the ocean, and when the temperature of the seawater falls below −1.8°C and ice crystals form. Within hours the millimetre-sized crystals aggregate, forming slicks of **grease ice** on the surface. As they rise through the water the crystals effectively 'harvest' particles, including the small planktonic organisms; some stick to the crystals, whereas others are simply trapped in the viscous, slushy ice layer. Many of the larger animals such as copepods, krill and fish avoid entrapment by actively swimming away.

Figure 12.4 Sea ice pancakes.

Figure 12.5 Closed ice and pressure ridge.

After a few hours of further freezing, the ice crystals accumulate to form loosely aggregated discs, **ice pancakes**, five to ten centimetres in diameter. These grow larger, becoming 20 to 50 cm thick 'super pancakes', several metres across. Wind and wave action raft the pancakes together, and often several end up lying on top of one another. They freeze together, and after one or two days a closed ice cover has formed, with an average thickness of <1.5 m, although old, rafted, deformed sea ice floes of >10 m are known.

The trapped sea ice organisms face considerable changes in their immediate physical and chemical surroundings as ice grows: Ice crystals are pure water, and as ice crystals form, the ions in the seawater are excluded and concentrated into brine solution. Compared to ice formed from fresh water, sea ice is not solid, but is laced by an intricate network of channels and pores filled with the brine. It is within this salty labyrinth, looking like a liquid-filled Swiss cheese or sponge, that the sea-ice organisms live.

Ability to withstand confined space, low temperatures and high salt concentrations are prerequisites for survival of an organism in sea ice. As the ice becomes colder, more ice crystals are produced from the brine, concentrating it even more. At the same time, space becomes limited: in sea ice at –2 °C the channels are large (millimetres/centimetres in diameter), and the brine only very slightly more concentrated than seawater. In contrast, at around –8 °C the channels are considerably reduced in number and narrower (micrometres in diameter), and the salinity of the brine becomes around 145. At –25 °C there are virtually no channels or pores, and the brines are so concentrated that some minerals are deposited as solid salts. By the late autumn/winter there is a gradient of temperature between the cold surface of the ice floes (down to –30 °C) and the bottom of the ice in contact with the seawater (–1.8 °C). Coupled with the temperature gradient there is a corresponding gradient of porosity and salinity; i.e. at the top, few channels and very concentrated brines,

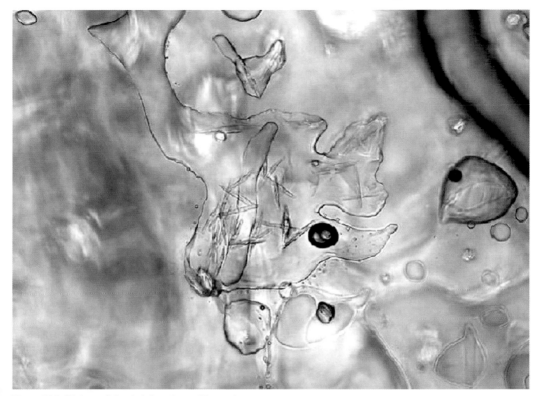

Figure 12.6 Diatoms living in brine channel in sea ice.

and at the bottom, large channels with normal seawater salinities.

The organisms living in ice have an intriguing biology in their own right, but are even more important in their role in the seasonal dynamics of the Antarctic and Arctic ecosystems. The abundant biomass within the ice provides a valuable food resource for organisms at a time when there is nothing else in the water to eat. Krill, for instance, are frequently observed grazing on the underside of ice floes in winter, and year to year fluctuations in the extent of sea ice have been linked to interannual variations in krill abundance.

12.2 Hydrothermal systems

In the 1970s marine scientists discovered the first hydrothermal vents in mid-ocean ridge regions of the Pacific Ocean. These are plumes of hot water, sometimes coloured black, that are generated from under the sea floor and can be at temperatures of >400 °C, although often the temperatures are much lower (between 10 and 100 °C). Although the temperature of the vent fluid is hot, it is rapidly cooled when it mixes with the deep seawaters at temperatures of <2 °C. The rapid cooling of the vent fluid can also result in minerals precipitating out rapidly, which can produce structures (>20 m

Figure 12.7 Black smoker taken from the Woods Hole Oceanographic Institute's *Alvin*.

high) reminiscent of chimneys, and this is one reason for their common name of 'smokers': black smokers have vent fluids containing black minerals, and white smokers transport less obviously coloured minerals. Many hydrothermal vents have been documented over the past 30 years at water depths between 2000 and 3000 m. Their formation is associated with the movement of tectonic plates and the formation of new oceanic crust, and so they are concentrated around the East Pacific Rise and Mid-Atlantic Ridge.

What has fascinated many since the discovery of these vent systems is that they harbour a rich diversity, and often very abundant communities of animals including worms, shrimps and clams. These are not feeding on the food falling from the photic zone above, but rather, at the base of the food chain for these communities are bacteria that use CO_2 and chemicals in the vent fluids for chemosynthesis (cf. photosynthesis, which needs light). For instance, many vent bacteria use hydrogen sulphide:

$$CO_2 + H_2S + O_2 + H_2O \Rightarrow [CH_2O] + H_2SO_4$$

These bacteria can be free-living in the water, or form dense mats on which animals graze. A rather specialised association between the bacteria and animals is that found in a group of worms known as the Vestimentifera. One of the most commonly described species is *Riftia pachyptila*, which are 1–2 m long worms that live in white tubes attached to the vent. Instead of a mouth or digestive system, they have a specialised organ called a **trophosome**, which contains large quantities of the chemosynthetic bacteria just described. These bacteria, which can be up to half the weight of the host worm, provide the carbon supply for the growth of the worm, which takes up the nutrients for itself and the bacteria by extending red plumes of tentacles into the surrounding water. Several of the clam species living in vent communities also host these bacteria in their gills and tissues in a similar symbiotic relationship.

Most of the animals living around vents are living on surfaces where the vent waters have been cooled to < 10 °C. A remarkable exception is the Pompeii worm (*Alvinella pompejana*), which can live in tubes close to the openings of the vent chimneys. Here they have been recorded living in waters around 80 °C and even surviving short spells at 105 °C. In one study the animals were recorded as having their tails at 70 °C and their heads at 20 °C, a gradient of 50 °C along their 10 cm long bodies.

Vents do not last particularly long and can stop flowing within 10 years of forming. Naturally when this happens it is a disaster for the animal community that has formed around the vent, especially the sessile ones, and without the supply of energy to fuel the bacteria the organisms will die. Therefore vent organisms must produce a lot of larval progeny

Figure 12.8 *Riftia* species and vent mussels at a hydrothermal vent.

that are able to travel in water currents in the deep sea with the chance of settling at another vent system.

12.3 Deep sea

Naturally the main characteristics of the deep sea are the high pressures encountered and the fact that it is in darkness. Pressure increases by 1 atmosphere per 10 m increase in water depth (1 atmosphere is the equivalent of 1kg cm^{-2}). So at the around 10,000 m the pressure is akin to the pressure applied by several elephants standing on a single dinner plate. But for the organisms living at any depth, this pressure is not an issue. It is only a problem when organisms are transported rapidly from high pressure at depth to the surface (as when being sampled), and typically the rapid pressure change causes the animals to become distorted and even explode on occasions. This does not mean that organisms do not migrate through significant pressure changes, and some species of lantern fish that live around 1,700 m during the day migrate upwards at night to depths of around 100 m or so to feed. Each up and downwards migration is a three hour journey.

Although no sunlight penetrates below 300 m, the deep sea is not a place without light, which explains why many deep-sea organisms have very well developed eyes. The light source (not strong enough for photosynthesis) is bio luminescence. Many groups of organisms in the sea, including bacteria, dinoflagellates, fish

Figure 12.9 Image of bioluminescence produced by dinoflagellates in the wake of a small boat near the Åland Islands in the Baltic Sea.

and squid, are able to create bioluminescence. This light can be produced by the organisms themselves, or by having bioluminescent bacteria contained in special structures. The light produced by most of these organisms is blue to green in colour (440 to 480 nm), and is caused by the breakdown of a chemical, luciferin, by the enzyme luficerase. There are a few animals that produce green or yellow light, and a small number that produce red.

The reasons for bioluminescence are various, but in the deep-sea organisms the light can be used to attract prey (angler fish) or mates (squid). Displays of light can also be used by some organisms to confuse or frighten predators. Others release decoys, including some that squirt out clouds of luminous particles to distract predators.

Bioluminescence is not restricted to the deep sea, and when it occurs on the surface of the ocean during a dinoflagellate bloom, can give a rather eerie glow to the ocean surface. Milky seas are well known in maritime folklore, and are likened to sailing upon a field of snow or gliding over the clouds, all under the darkness of a moonless night. In recent years satellite experts have tied up the reports of such events in ships' logs with satellite images, and have confirmed that bioluminescent events spanning many thousands of square kilometres do occur. It is thought that they are caused by massive accumulations of bioluminescent bacteria species growing within dying phytoplankton blooms.

Many organisms from the deep sea are certain to hit the news headlines when they are found accidentally on the surface of the sea. The most notable of these are the Giant Squid (*Architeuthis dux*) or Colossal Squid (*Mesonychoteuthis hamiltoni*) which can be over 15 m in length. Contrary to myth, they do not attack ships, and are no match for a sperm whale; however, large specimens will need to feed on over 50 kg of food per day, and so are major predators of the deep. This is one of the reasons that they have the largest eyes known in biology, reaching diameters of up to 30 cm.

Gigantism is not confined to the squid, but there are many cases of deep-sea organisms (amphipods, isopods and sea spiders amongst others) being larger than similar species from warmer waters. It does seem as though part of the reason is that many of organisms may be long-lived, and how these organisms live so long is the subject of intensive studies (partially in quest of any clues they may give about longevity/ageing in humans). One possible theory is that their longevity is linked to the low cell metabolism at cold temperatures. Normal metabolism produces reactive oxygen products such as oxygen radicals and hydrogen peroxide. These potentially toxic products are thought to be influential in the ageing process. At the low temperatures in the deep sea and in polar regions, cellular respiration rates are low, and so the production of these harmful substances will also be low, enabling organisms to live longer. It is not just invertebrates that can reach an advanced age in the deep sea; there are also long-lived fish species, such as the Orange Roughy (*Hoplostethus atlanticus*), which can live for up to at least 149 years, not reaching sexual maturity until its mid-20s.

13 Changing oceans

There is no doubt that one of the major environmental debates facing the human population is the potential effects of global climate change, sometimes referred to as global warming. Records from ice cores taken from the ice sheets in Greenland and the Antarctic show that climate change has occurred many times over hundreds of thousands of years, but it is the global climate changes since the 1850s that are foremost in the general public's thinking, and that drive global environmental policy agendas. Sophisticated climate models are being employed to predict how much global temperatures will increase, and the current predictions range between 1.8 °C and 4 °C by about 2100. In this chapter three different (although often related) large-scale environmental changes will be introduced, and all have large impacts from the scale of an organism through to the entire globe (Figure 13.1).

13.1 Global climate change

The consequences of global climate change for the oceans are many and varied, and tied up with complex interactions between air temperature, weather patterns, and ocean circulation. Here the effects of some of these on the Arctic Ocean will be used to illustrate the changes that are taking place in the oceans.

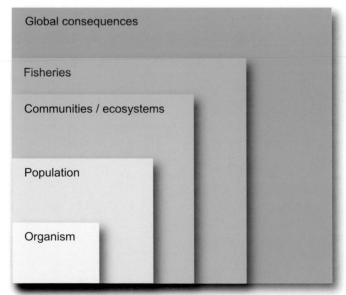

Figure 13.1 The effect of large-scale environmental change impacts on many scales. On the organism scale it includes change in primary production, growth rate, reproduction, dispersal. On the population scale it includes change in population growth, species distribution and abundance. At the level of community/ ecosystems change induces alteration to trophic interactions, biodiversity and community structure. Impacts at the scale of fisheries include changes in catches and fisheries management policies. On the global scale, human population growth and migration are impacted through both food and energy supply.

Global consequences

Fisheries

Communities / ecosystems

Population

Organism

Since the early 1980s, average air temperatures at latitudes higher than 60 °N have increased by about 0.5 to 0.9 °C per decade. In general, springs are warmer and coming earlier in the year. A direct consequence of this warming trend is that some areas of the Greenland ice sheet have increased their rate of melting, and there has been increased melting of large glaciers in other Arctic regions. This melting has led, in part, to the 10 to 20 cm increase in sea level since 1900, or the current estimate of sea level rise of 3 mm per year.

NB Sea level is rising with increased input of fresh water from glaciers and ice sheets. This is not to do with melting pack ice (ice formed from frozen seawater) since sea ice is less dense than seawater, and when it floats on the ocean surface it only displaces its own weight of seawater (see Chapter 2). If all the sea ice in the earth melted, the melt water would simply replace the volume previously occupied by the ice. That is why there are no sea level changes with the annual formation and melting of the millions of square kilometres of sea ice each year.

It is not just increasing temperature itself that is causing the dramatic changes being recorded. Over the past few decades it has become clear that many oceanographic trends in the northern hemisphere, and the sea ice dynamics, are closely linked to the North Atlantic Oscillation (NAO) and the closely related Arctic Oscillation (AO). The NAO index is a measure of the atmospheric pressure differences between the region of Greenland/Iceland and the subtropical central north Atlantic in the Azores. The NAO index is defined as the difference between the Icelandic low and the Azores high in winter (December to March).

When the NAO index is positive there is a strong Icelandic low and Azores high pressure, with a corresponding strong North–South pressure gradient. During negative NAOs the pressure gradient is weak, with an Icelandic high and Azores low. The warming in the Arctic has been gradual over the past 100 years. However, the pace of this warming over the past 20 years has been increasing at a rate 8 times higher than the longer 100 year trend, and the rapid warming trends are thought to be associated with increasing positive phase in the NAO index (Figure 13.2).

The most dramatic large-scale changes to be recorded in any marine system in the past 30 years are the changes in yearly minimum sea ice cover in the Arctic, which has been decreasing at a rate of about 6.8% per decade. The consequence is an annual decrease in mean annual sea ice cover of around 4% per decade. It is not only the extent that is reducing, but also the thickness of the ice: between the 1960s and 1990s there has been a decrease in sea ice thickness of more than 2 m in the central Arctic. These sorts of findings have led several researchers to conclude that there will be no summer sea ice in the Arctic after the 2050s. This would mean that although sea ice would form in winter, it would all melt in the following spring/summer, and there would be no thick pack ice, which was characteristic of the Arctic in the past. Clearly such dramatic changes in the seasonal dynamics of the physical structure of a whole ocean basin will have profound implications for the behaviour of the oceanography of the Arctic Ocean, and the seasonal dynamics of the ecosystem that has evolved around a sea ice cover lasting throughout the year.

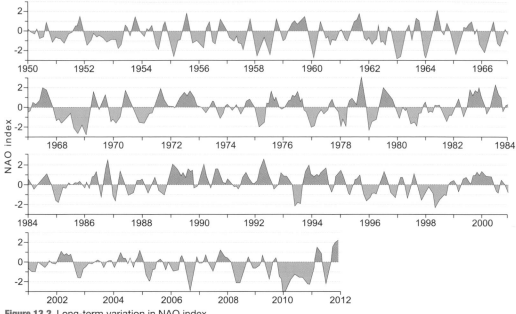

Figure 13.2 Long-term variation in NAO index.

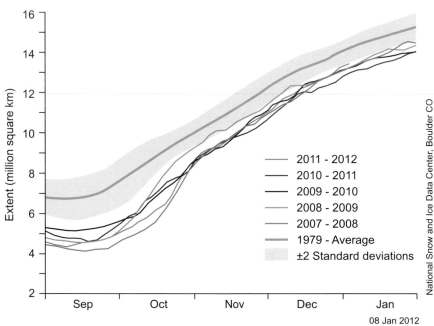

Figure 13.3 Changes in sea ice extent in the Arctic. The ice conditions since 2007 have been consistently lower than that recorded by satellite measurements made in the period 1979–2000.

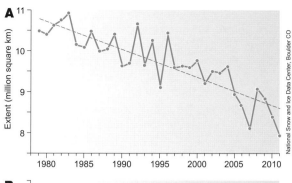

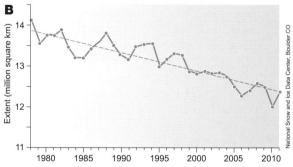

National Snow and Ice Data Center, Boulder CO

Figure 13.4 Detailed picture of changes in monthly ice extent between (**A**) July and (**B**) December between 1979 and 2011. The July trend shows a decrease of 6.8% per decade and the December trend a decrease of 3.5% per decade.

Figure 13.5 The sea ice and ice shelf in this image have high albedo (are highly reflective) whereas the seawater has a low albedo (absorbs energy)

Ice is a very good reflector of incident solar radiation, whereas seawater is a comparatively good absorber of heat energy from solar radiation. Ice and snow are said to have a high **albedo**, whereas the dark water has a low albedo. When ice melts, the albedo of the remaining ice is reduced, and therefore more energy can be absorbed. This in turn will increase the rate of melting. This is termed a *positive albedo feedback loop* where the absorption of heat energy leads eventually to an even greater absorption of energy. If there is an increased melting of sea ice in the Arctic, the reduced albedo will induce further warming of the surface waters, with thinner ice resulting in accelerated ice melt.

13.2 Ocean acidification

There are two contrasting issues for the carbon chemistry of the oceans in the future warming environment that is being predicted by climate scientists:

1   As oceans warm, they will absorb less CO_2 from the atmosphere, since the solubility of CO_2 is less in warmer water.

2   As more CO_2 is absorbed by the oceans due to the rising concentrations in the atmosphere, there will be reduction in the pH of the seawater, a process commonly referred to as *ocean acidification*.

Since the Industrial Revolution in the 1700s it is estimated that the average pH of the oceans has dropped from 8.16 to 8.05. It is predicted that by 2100 the global oceans may have undergone a further decrease of about 0.4 pH units (Figure 13.6). Such changes are greater than any pH change thought to have happened at any time over the past 300 million years.

At the moment the surface of the oceans is saturated in the various mineral forms of calcium carbonate, and these do not dissolve at these depths, whereas they do dissolve

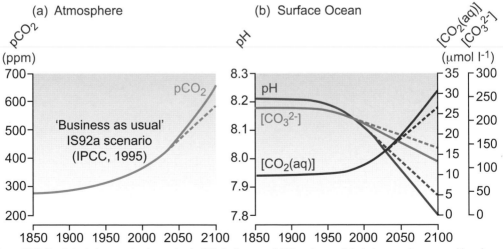

Figure 13.6 Predicted changes in atmospheric CO_2 to the year 2100 and associated changes in ocean pH and carbon chemistry (cf. Chapter 9).

in deeper waters that are under-saturated in these minerals. Because of the lowering of pH of ocean waters, and the associated reduction of carbonate ions, the surface waters are becoming less saturated in calcium carbonate. It is predicted that in the next 150 years the carbonate-saturated surface waters in some regions, especially cold Arctic and Antarctic waters, may disappear altogether. It is intriguing that in the Arctic, the organisms and ecosystems are not only being impacted by massive changes in sea ice dynamics, but also undergoing change due to pH.

There are many groups of organisms that use calcium carbonate in their body skeletons and/or external structures such as shells. These include the corals and molluscs, but also echinoderms, crustaceans, some seaweeds, foraminifers and calcareous phytoplankton such as the coccolithophores (Chapter 8). Naturally, anything that reduces the availability of carbonates for building these structures will impact on the organisms' ability to grow and survive.

Currently there are many long-term experiments being conducted by marine biologists to see what the impacts will be, but there are now many reports of shells/skeletal structures being thinner/weaker when the organisms are kept in lower pH waters. Obviously if such effects occur in nature there will be significant impact on the populations of calcifying species. It is not simply a matter of taking adults and observing the changes that occur, but it is important to see the effects of lowered pH on all stages of life history, since some stages may be more vulnerable than others (e.g. larvae). This has been highlighted in experiments with fish, where the eggs were more vulnerable to high-CO_2 induced

mortality than the larval stages. Increased ocean acidification has also been shown to cause considerable tissue damage to larvae of the Atlantic cod (*Gadus morhua*), which clearly will impact on the stocks of these fish.

13.3 Eutrophication

As described in chapters 9 to 11, the growth of phytoplankton is limited by light, inorganic nutrients, and whether or not it is eaten by zooplankton. If there is a plentiful supply of the first two factors and no grazing, then the amount of algal biomass that grows in a water body can be very great indeed (Figure 13.7).

Figure 13.7 This shallow rock pool has been fertilised by the guano from birds and has lots of light, so the algal growth has been prolific. Compare the colour of the pool with that of the sea in the background.

Figure 13.8 River discharging into coastal waters that can transport the nutrients in run-off from agricultural land.

Over the past 100 years there has been a dramatic increase in the amount of nitrogen and phosphorus that has been applied to land to increase crop production, and this trend has sharpened especially in the past 20 years. Ultimately, much of this nitrogen and phosphorus runs off into rivers, and eventually through estuaries into coastal waters. The resulting increase in phytoplankton growth is well documented and is called **eutrophication**.

Although human-induced nutrient enrichment is the most common cause, eutrophication can result from a range of processes that increase the rate of supply of organic matter to an ecosystem, and is not confined to the increase of nutrient supply alone. So something that increases the light availability for photosynthesis, such as a reduction in the load of suspended mineral particles, could result in increased algal growth, and therefore cause eutrophication. A change in the residence time of water in an estuary could also lead to eutrophication through increased algal production. Eutrophication is generally perceived as a detrimental process, which can cause multiple harmful effects on an ecosystem stemming from increased production of organic matter within the system. Harmful effects can be divided into direct effects related to phytoplankton growth (e.g. increased biomass, algal blooms) and indirect effects (reduced biodiversity, increased biomass of opportunistic species, oxygen deficiency on the sea floor) caused by excess organic matter in the system. Eutrophication is typically a human-induced problem, and it is important to stress that it can also be reversed by proper management. There are also instances when low levels of eutrophication can even be considered as being a positive state for increasing the productivity of a water body or coastal system.

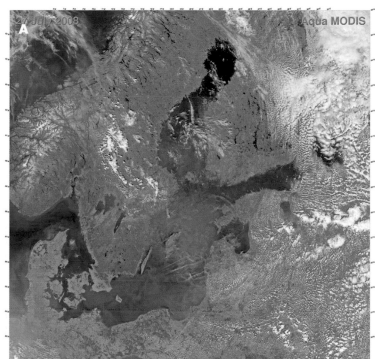

Figure 13.9 (**A**) Algal bloom in the Baltic Sea, which has been subjected to eutrophication for several decades. (**B**) Surface water view of Baltic Sea cyanobacteria bloom in eutrophic coastal waters.

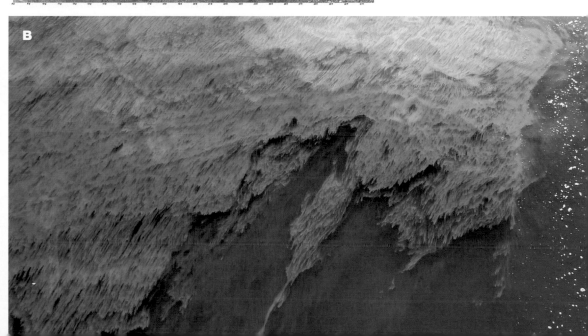

Where eutrophication does occur on a large scale, and over sustained periods of time, the consequences for the ecosystem can be dramatic. This is especially true of enclosed, or semi-enclosed and permanently stratified water bodies such as the Baltic Sea. The increases in phytoplankton growth in the Baltic have been associated with increased turbidity of the water reducing the light available for growth of submerged seaweeds and plants. Also the diversity of phytoplankton species has changed, with different species now dominating the communities than in the past. However, the most dramatic change has resulted from the sinking of the increased phytoplankton blooms to the sea floor. The Baltic Sea has a permanent salinity stratification, which slows the supply of oxygen to near-bottom waters. The increased breakdown of organic matter has led to such an increase in respiration that the oxygen concentrations close to the seabed have become very low in places, and in some regions so low that benthic organisms have been killed.

14 Sampling the oceans

Edward Forbes (1815–1854) is often credited with being the founding father of oceanography. However, it was his successor to the Chair of Natural History at Edinburgh University, Charles Wyville Thomson (1830–1882), who had the vision for, and was the leader of, the Challenger Expedition (1872–1876), a remarkable voyage visiting all of the world's oceans except the Arctic. This was arguably the first multidisciplinary oceanographic research expedition, and the resulting 50 volumes of final reports laid the foundations for modern day oceanography.

Modern day oceanographic cruises are just as much voyages of discovery. Marine scientists lower instruments into the ocean to measure change in water properties with depth. They deploy nets and water bottles and seabed grabs. Some instruments, such as current meters, can be left on moorings to record data for weeks, months or years, and increasingly our information about the ocean is gathered from satellites orbiting the earth.

14.1 Research vessels

Many nations operate a fleet of purpose-built research vessels. An example of a medium-sized ship, suitable for research cruises in shelf seas, is shown in Figure 14.1B. Scientists work in 'watches' around the clock, sleeping and eating on board, and research cruises can last anything from just a few days up to two or three months In more remote regions of the world's oceans. For deep-ocean work, larger ships would be used (Figure 14.1A) and small survey boats (Figure 14.1C) are ideal for coastal work.

It is important to realise that research vessels generally operate for most of the year, and teams of scientists join the vessel only for periods of time. This means that the ship's laboratories are largely empty and need to be set up at the beginning of each journey. This can be an arduous task, since space is of course limited, and there is often competition to get enough room. It is also important to have done your packing beforehand very carefully. There are no shops in the middle of the North Atlantic or Southern Ocean, and once the ship leaves port it could be several months before a harbour is visited again. It is remarkable how, in just a matter of hours, an empty room can be transformed into a high quality analytical laboratory (Figure 14.2), albeit with everything tied down to stop it falling off the benches in rough weather. It is even more remarkable because the scientists have to construct their new laboratories while still having to find their *sea legs* and maybe suffering from sea sickness, which is an occupational hazard for many.

14.2 Measuring water column structure

The workhorse of any oceanographic cruise is the Conductivity–Temperature–Depth probe (CTD; *see* Figure 14.2). This makes the basic measurements of water temperature and salinity at depth in the ocean. The CTD is lowered on a steel wire that has an electrical

Figure 14.1 Four types of research vessel: (**A**) large and suitable for deep-ocean work; (**B**) medium and suitable for shelf seas; (**C**) small and suitable for coastal and estuary studies; (**D**) improvised use of a fishing boat in the Philippines.

cable inside. The data from the instruments is transmitted up the cable to computers on board the ship. A CTD measures the electrical conductivity and temperature of the water and from these two properties, salinity can be calculated. Depth is determined from a pressure sensor in the CTD. Other instruments will normally be attached to the CTD 'frame'. These are likely to include a fluorometer to measure chlorophyll fluorescence, and a transmissometer or optical backscatter sensor for water turbidity.

Figure 14.3A shows a CTD profile in a water depth of about 50 metres in the Irish Sea. The water column is well mixed: the temperature in blue, salinity in red, transmittance (a measure of water turbidity) in white and fluorescence in green are all the same from top to bottom

Figure 14.2 Before and after setting up a laboratory on board a large research vessel (A and B). This contrasts with the improvised laboratory (C) on deck a small boat in Greenland where working temperatures never got above 5°C.

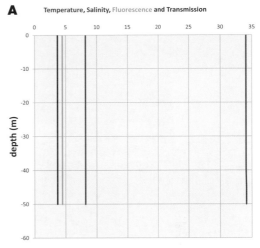

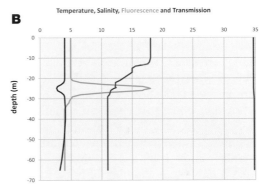

Figure 14.3 Results of a CTD profile in (**A**) vertically mixed water and **B**) stratified water. NB The scale for temperature is in degrees centigrade, and salinity in units of salinity. However, the fluorescence (measure of chlorophyll *a*) and transmission are in arbitrary units of voltage that are converted to actual values by the comparison of the readings against certified standards.

of the profile. In a different part of the Irish Sea (Figure 14.3B) the profile is very different. There is a thermocline from the surface down to about 30 metres. The white transmittance profile shows clear water lying on top of more turbid water, and the green fluorescence profile shows a fluorescence peak around the bottom of the thermocline.

Water samples can be collected from depth in a ring of bottles, called a rosette sampler,

Figure 14.4 A small undulating frame for towing behind a research vessel under way. The frame can be fitted with instruments such as a CTD and the pitch of the 'wings' adjusted such that the instrument 'flies' up and down in the wake of the ship.

surrounding the CTD (*see* Chapter 1). The bottles are cylinders with caps at the top and bottom. When the CTD goes into the water, the caps are open. The bottles are fired at chosen depths by sending an electrical signal down the CTD wire. This closes the caps, trapping a sample of water in each bottle. A number of samples can be collected at different depths, limited only by the number of bottles in the rosette.

Once back on board the water from the bottles can be decanted carefully for a range of analyses, and routinely these include salinity (to calibrate the CTD sensor), nutrients, dissolved gases, numbers of organisms and particles. The bottles are fine for collecting bacteria and phytoplankton, but other equipment has to be used for collecting zooplankton, larvae and nekton. If the water in the bottles is to be used for the analysis of trace elements such as dissolved iron, specially constructed and cleaned bottles have to be used, often coated with Teflon, and certainly not including steel pins, hinges etc.

In shallow water, a CTD profile may only take a few minutes, but in the ocean, in water kilometres deep, it can take several hours. Often a grid of stations will be worked so that the water column structure can be measured both vertically and horizontally. A CTD can also be towed behind the ship on a specially designed 'towed undulator' frame (Figure 14.4). The towed undulator is fitted with controllable fins so that it can be made to dive and surface behind the ship, so giving a continuous up and down profile of measurements in the ship's wake.

14.3 Measuring water flow

An example of a recording current meter is shown in Figure 14.5A. The current meter has a vane which makes it line up with the current; a compass in the body of the instrument records the direction of the flow. The impellor turns in the current and the number of turns in a set interval of time is proportional to the current speed. A current meter like this one is battery operated. It is deployed at sea on a mooring at a fixed location, and so gives an **Eulerian** measurement of current. The traditional type of oceanographic mooring is U-shaped (Figure 14.5B). A sub-surface buoy, of sufficient size to support the instruments and the line holding them, is paid out over the side of the ship. The line (or riser) with the instruments attached is then paid out. There is no tension in the line as the instruments go into the sea, and so they can be lifted over the side by hand. The first ground weight is then lowered to the seabed by the ship's winch and (if the lengths of line have been judged correctly) the sub-surface buoy will disappear below the water surface. The ship then steams slowly ahead while the ground line is paid out. Finally, the second ground weight, second riser and surface marker buoy go over the side. Recovery is the reverse of deployment, and so again there is no tension on the line when the instruments are brought aboard. U-moorings have the advantage that ships can pass over the sub-surface buoy without dragging the instruments away. An increasingly popular alternative is to use a single riser with an acoustically-operated release mechanism just below the ground weight. When the release is activated (by a coded signal) the sub-surface buoy is released and floats to the surface with the instruments dangling below.

A number of current meters can be placed on the mooring at different heights above the seabed to measure the profile of the current – that is, the way the velocity changes with depth. Alternatively, a single acoustic Doppler current profiler can be used. Figure

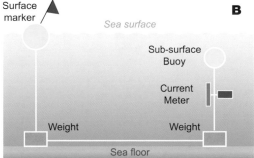

Figure 14.5 (A) A recording current meter and **(B)** a schematic of a U-mooring, on which a recording current meter can be deployed to make continuous measurements of current speed and direction.

14.6 shows an ADCP fixed in a frame which will be lowered to the seabed. This type of current meter sends pings of sound into the water. Some of the sound energy is scattered by particles in the water and returns to the ADCP as an echo. If the water is moving, the

Figure 14.6 (A) An Acoustic Doppler Current Profiler (ADCP) which contains the acoustic transducers. **(B)** An ADCP in a frame ready for deployment on the seabed. The large yellow buoy to the left is the surface marker buoy.

echo has a slightly different frequency (the Doppler shift) to the transmitted ping. The velocity of the water can be calculated from the Doppler shift. By collecting echoes from different heights, the profile of velocity above

the bed-mounted ADCP can be constructed. Research vessels can also be equipped with hull-mounted downward-looking ADCPs to measure the velocity profile below the ship as she steams along. The ship's speed needs to be subtracted from these measurements to give the true water velocity. If the water is shallow enough so that an echo is received from the sea floor, the ship's speed can be calculated from the Doppler shift in this echo.

14.4 Satellite remote sensing

The oceans are large, and research vessels can only cover a small part of their area during a single cruise. In contrast, a satellite orbiting the earth can cover the entire ocean every day. Instruments aboard a satellite can make measurements of some surface water properties from a height of several hundred kilometres above the water – a procedure called remote sensing. In *passive* remote sensing, the satellite measures the electromagnetic radiation (mostly in the form of visible light and infra-red radiation) leaving the water surface. Figure 14.7 shows a satellite image of the seas of north-west Europe in the visible part of the spectrum. The regions of bright, coloured water in the English Channel and Irish Sea are caused by particles suspended near the surface. A lot has been learned about the behaviour of suspended matter in shelf seas using images such as this. We have already seen in Chapter 6 how infra-red images can be used to study the surface temperature of seawater and can be used to detect fronts. It is also possible to use passive remote sensing at longer wavelengths to measure the surface salinity of open ocean waters.

Active remote sensing involves sending a pulse of electromagnetic energy from the

14.5 Sampling difficult places to reach

Sampling the organisms of the deep is a difficult thing to do, and has been compared to sampling a rainforest from a plane being flown 2 km above the clouds, dangling a net below the plane and pulling it through the forest canopy. How would you ever reconstruct a decent idea of the rainforest from the bits and pieces that would collect in the net? However, that is exactly what most of the deep-sea research has had until very recently. Just trawling a net at 5000 m actually requires having a cable of twice the length, considering the angle that needs to be employed between the ship and net for effective towing: 10 km of heavy steel cable has to be deployed from a huge cable drum and can be carried only on the largest of research vessels.

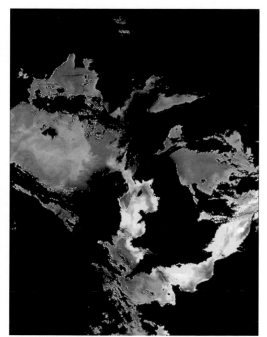

Figure 14.7 A satellite colour composite image of the seas around the British Isles. Bright coloured water in the Irish Sea and English Channel is caused by suspended particles scattering light. Land and clouds are coloured black

satellite down to the sea surface and measuring the properties of the signal that bounces back. Energy at radar wavelengths (a few centimetres) are most commonly used. The time of flight of the radar pulse gives the distance between the satellite and the water surface. If the satellite height is determined independently (using ground stations) the undulations of the ocean surface can be measured. These *altimetry* measurements are valuable for studying ocean tides, currents (because of the geostrophic balance) and waves. The exact nature of the radar echo depends on the roughness of the sea surface, and radar images of surface roughness can be produced.

Much of what we know about the deep sea, especially systems like hydrothermal vents, has come from the development of manned submersibles that can withstand the large pressures of the deep ocean. Possibly the most famous of these is the USA's *Alvin* that can dive routinely to depths of 4500 m. There are a number of such DVSs (Deep Submergence Vehicle), many of which can go down to 6000 m or more. These are equipped with a range of robotic arms for collecting samples and other sophisticated sampling devices for water, organisms and deep-sea sediment samples. The deepest manned dive was a two-man mission in the *Trieste* in 1960 to a depth of about 10,900 m. DVSs require considerable investment, since the infrastructure in ships and personnel to support the missions is considerable.

Unmanned vehicles, or remotely operated vehicles (ROVs) are of course a less demanding technology than sending humans down to the

Figure 14.8 (**A**) The DVS *Alvin* being launched, and (**B**) closer view showing the complex array of sampling devices, bottles, sensors and camera equipment.

depths. These can carry cameras, videos, water samplers and even robotic arms to collect samples. The depth limitation of an ROV is of course the length of cable that attaches the vehicle to the operator on ship. However, vehicles like *Nereus*, operated by Woods Hole Oceanographic Institution in the USA, is able to be operated as an ROV, but can also be untethered from the cable for wider-scale studies, thereby operating as an Autonomous Underwater Vehicle (AUV). AUVs are programmed to fulfil a sampling programme/ survey and then surface at a specified location.

Another way of measuring what is going on at the ocean floor is to send down *landers*, which are structures designed to sit on the sediment surface and make measurements or take water samples at programmed time intervals. Often landers, as well as containing probes, sensors and sampling devices, also have video and time-lapse cameras to record the organisms present at the site. Some landers are dedicated to just taking images, often baited with food to attract organisms. The lander is retrieved by sending an acoustic signal to release it from the weights holding it on the bottom, after which it then floats to the surface and is retrieved. Landers are routinely deployed for periods of days, through to several months.

Sediment traps are frequently deployed to catch the flux of particles through the water column. They are typically constructed in the form of an inverted cone, so that a large surface area of the structure collects the material falling down, and this is concentrated in the narrow part of the cone, where it is collected in bottles. Each sediment trap will be equipped with a number of sample bottles, and the device programmed so that each bottle will

Figure 14.9 ROV being deployed in the Arctic. NB the thick cable coming out of the top, which connects it with the operator on board the ship.

Figure 14.10 (A) Autosub which is an AUV, that gets some curious looks from the resident animals such as this whale in the Southern Ocean (B).

Figure 14.11 A benthic lander about to be deployed.

collect material for a defined period of time. In this way a sediment trap could be left in the water for a year or even more, and the flux of sinking material for each month collected in a single sample bottle. The bottles have a poison in them so that the material is preserved and doesn't decay before analysis once the trap is retrieved. Often traps are hung at different depths on a single mooring. Like the lander, when the deployment is finished an acoustic signal is sent to release the array of traps from the weights holding the line in position, and they float to the surface.

As exciting as these technologies are, they are still restricted to the places where ships travel, and therefore a large part of the world's oceans will be sampled only infrequently, if at all. What are needed are roaming platforms

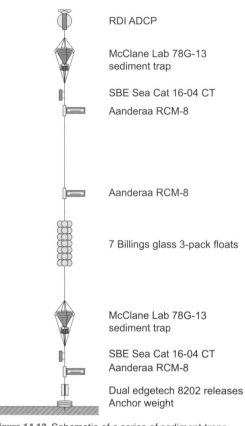

RDI ADCP

McClane Lab 78G-13
sediment trap

SBE Sea Cat 16-04 CT
Aanderaa RCM-8

Aanderaa RCM-8

7 Billings glass 3-pack floats

McClane Lab 78G-13
sediment trap

SBE Sea Cat 16-04 CT
Aanderaa RCM-8

Dual edgetech 8202 releases
Anchor weight

Figure 14.12 Sediment trap being deployed in the Arctic. NB the green and orange current meter beneath it that is important for recording the current speed and direction during the sediment trap deployment.

Figure 14.13 Schematic of a series of sediment traps and current meters deployed in the deep ocean.

that are able to measure physical, biological, and chemical parameters across depths and for long transects. A whole range of AUVs, gliders, and floaters have been under development since the late 1990s (Figure 14.14). These battery-powered devices are designed to 'roam' pre-programmed tracks over large regions of the ocean, collecting data. Periodically these devices surface to send the collated data to satellites, which then transfer the data to base stations, from where scientists can download the information. To date, many of these devices carry sensors for salinity, temperature, and pressure, among others. However, as technology progresses other sensors such as fluorescent and nutrient sensors will be routinely deployed on such platforms.

14.6 Sampling the biology

Collecting water from the rosette bottles deployed with the CTD is the routine way of sampling the phytoplankton and bacteria.

Figure 14.14 Seagliders from the University of East Anglia, UK being deployed in the northern Weddell Sea (Southern Ocean) in January 2012 from the RRS James Clark Ross. Three Seagliders were deployed as part of the GENTOO project (Gliders: Excellent New Tools for Observing the Ocean) in order to study the dynamics of the Antarctic Slope Front and its role in maintaining the krill-based ecosystem.

However, zooplankton (even small ones) and fish are more typically sampled using nets. These vary in shape and size and can either be trawled behind a moving ship or vertically through the water column. The latter method is vital if you want to know the vertical distribution of the organisms in the water, and devices are deployed to close the nets at desired depths as the nets are hauled up vertically. In some instances nets are hauled singly, but frequently devices holding multiple nets are cast at a single time. Naturally with nets, the size of the mesh determines the size of organisms caught, and as a rough guide a mesh size of 75% of the width of the smallest organisms requiring to be sampled is needed for quantitative sampling; e.g. for copepod samples, nets with mesh sizes of around 200 to 500 μm are frequently used.

Figure 14.15 Different types of vertical (**A** and **B**) and horizontally towed nets (**C**).

The organisms collected in the nets are obviously easy to collect for sorting and counting when the net is retrieved. However, for the smaller suspended particles, phytoplankton and bacteria, it is not so straightforward. These generally occur at such low concentrations that they have to be concentrated, which is done by either letting the samples settle, or more usually by filtering through filters with defined pore sizes. The chemists interested in the dissolved constituents of the water also have to filter out the particulate material before they can analyse the water. For many scientists on a ship, a very large amount of time can be spent laboriously filtering water, since it has to be done slowly

Figure 14.16 Seals with sensors (position, depth, temperature and salinity sensors) attached with glue to the fur on their backs. The sensors are attached while the seals are sedated (in the tent in the background) and these two are just waking up. When the seal sheds its fur the tags will be lost.

under gentle vacuum pressure so as not to damage the organisms.

An alternative to using nets for fish and zooplankton is to use acoustic devices. These devices transmit regular pulses of sound of a single frequency (tens to hundreds of kiloherz), which hit a target, and the sound reflected from the organisms is measured by a receiver. Fish that have swim bladders filled with gas are particularly good reflectors; however, acoustic techniques are effective for measuring many zooplankton species, especially those that form dense layers in the water, such as krill. Acoustic devices left on moorings at one spot are particularly useful for recording diurnal migrations of zooplankton and fish.

Satellite technology can be used for tracking the migration and movements of large animals in oceans. Tags, which send data back to researchers via satellite links, have been attached to whales, seals, turtles, and birds, and the position of the animal is recorded for as long as the transmitter functions or stays attached to the animal. As this technology advances, the sensors included on the tags have expanded to include depth sensors for recording the diving behaviour of the animals; mini CTDs to record salinity and temperature of the waters; devices to record when an animal opens its mouth under water to show at what depth it is feeding; mini cameras to record an animal-based perspective of what it is like under water.

Mammals can even be used to collect oceanographic information from regions that cannot be accessed by ships, such as beneath ice. Whales or seals can spend considerable time under ice or at ice edges, but need to surface regularly to breathe, and so are ideal candidates for such *biological AUVs*. In one study two Beluga whales were equipped with devices to measure depth, salinity and temperature in an Arctic ice-covered fjord. The data was logged as the whales dived to depths of 180 m, and when the whales surfaced to breathe the data was automatically transmitted to scientists via satellites. Although the sensor on one of the whales failed, data from the other was returned for 63 days, enabling the distribution of water masses under the ice to be described over an area of 8000 km². Such information would never have been possible with a ship.

14.7 The long term

As oceanographers contemplate the effects of long-term change in the oceans, it is paramount that long-term data sets of comprehensive information are available. There are two ways of collecting such information: keep going to the same place on a regular basis or deploying moored instruments to collect data and or samples.

One of the most imaginative ways of getting such data is the use of cargo or ferry ships, which routinely travel along the same courses. Automatic water samplers and/or sensors can be installed on these so that detailed information can be collected over long periods of time. A pioneer who recognised the importance of such surveying was Sir Alister Hardy (1896–1985). He invented the continuous plankton recorder (CPR), a device towed by a ship at a depth of about 10 m, which collects the plankton onto a silk ribbon. The ribbons are returned to the Sir Alister Hardy Foundation for Ocean Science in Plymouth, UK (http://www.sahfos.ac.uk/) where the plankton on the ribbon are analysed to measure phytoplankton and zooplankton species abundance and colour as a measure for the density of plankton. The first tow was in 1931 from Hull, UK to Bremen, Germany, and since then more than 275 vessels have been used to tow CPRs, and the North Sea and North Atlantic have been routinely sampled on a series of shipping routes since the 1940s to the present day. In a single year there can be over 380 tows covering distances of more than 130,000 nautical miles.

Two long-term oceanographic programmes were inspired in their conception, and their implementation has been fundamental to our understanding of seasonal changes in physics and biogeochemical processes in the oceans. These are the monthly samplings taken at a deep water in the Pacific, in the *Hawaii Ocean Time Series* (HOTS, http://hahana.soest.hawaii.edu/hot/hot_jgofs.html), and a sister project in the Atlantic, the *Bermuda Atlantic Time Series* (BATS, http://bats.bios.edu/index.html). Scientists have been making monthly measurements at these two stations since 1988 to include temperature, salinity, major inorganic nutrients, gas content of the water, pH, chlorophyll, bacteria and phytoplankton productivity, among others. They also have long-term moorings at the two stations. These two remarkable time series are a considerable investment of both resources and scientific effort. However, the information on the inter-annual variability has been invaluable in unraveling our knowledge about the changes occurring due to short-term variability compared with long-term change.

Figure 14.17 Despite all the progress made over the past 200 years by tens of thousands of dedicated oceanographers from around the world, we are routinely surprised by new discoveries from the depths. Exploring the oceans can be demanding, and taxes human ingenuity to the limits. However, the lure to take to the sea is a powerful one, and for those of us who have had the privilege to dedicate our careers to the study of oceanography, the rewards are immense.

Glossary

abyssopelagic [67]**:** the zone in the ocean ranging from about 4000 m down to 6000 m water depth.

acoustic Doppler techniques [28]**:** these techniques are used to make remote measurements of water currents. A pulse of sound transmitted into the water is scattered off suspended particles and the echo is received back at the transmitter. The time delay between transmission and echo gives the distance to the scattering material. The Doppler frequency shift between transmitted, and echo pulses, give the speed at which the suspended particles (and hence the water) are moving. By processing echoes from different distances, a profile of water current velocity can be constructed.

albedo [124]**:** the fraction of the incident shortwave radiation that is reflected from a surface. Snow and ice have high albedo and dark surfaces such as open ocean have a low albedo.

amphidromic system [44]**:** describes a wave created in a basin on a rotating Earth. The wave travels around the edge of the basin, in an anti-clockwise sense in the northern hemisphere, around a point in the centre called the **amphidromic point**. At the amphidromic point there is no vertical wave motion, but maximum current.

archaea [70]**:** one of the three major domains of life (the other two are the Bacteria and the Eukaryotes).

Archimedes' Principle [11]**:** states that when an object is placed in water it experiences an upthrust (or upward force) equal to the weight of water it has displaced. A special case of Archimedes' principle is that a floating body displaces its own weight of water.

aspect ratio [1]**:** the ratio of two dimensions of an object, for example the width and depth of an ocean.

azoic hypothesis [111]**:** this was the hypothesis proposed in the early to mid-1800s that nothing lived in the deep sea. This is clearly not true.

barycentre [00]**:** the centre of mass of an object, or series of objects. The barycentre of the Earth and Moon together is an important point in tidal studies. Both Earth and Moon revolve about their barycentre once a month.

bathypelagic [67]**:** the zone in the ocean ranging from about 1000 m down to 4000 m water depth.

benthic [67]**:** a term given to organisms living on the sea/ocean floor (the benthos).

buoyancy [11]**:** the upward force on an object placed in water. *See* **Archimedes' Principle**.

caballing [14]**:** when two cold water masses mix, the density of the mixture can be greater than the density of the original water masses. This effect, called caballing, is produced by the non-linear dependency of the **density** of seawater on temperature and salinity and is thought to contribute to the formation of the densest and deepest water in the ocean.

chart datum [37]**:** the lowest level the tide will ever reach at a given location. Depths on UK Admiralty charts are expressed relative to chart datum.

ciliates [72]**:** a group of protozoans characterised by having numerous hair-like organelles (cilia) which they use for movement and collecting food particles.

compensation irradiance [95]**:** irradiance above which algae produce more oxygen by photosynthesis than they consume by respiration.

Coriolis effect [31]**:** this is the term given to the apparent deflection of moving objects on a rotating earth. Viewed from the turning earth, objects actually moving in a perfectly straight line when viewed from space appear to bend to the right in the northern hemisphere and to the left in the southern hemisphere.

cyanobacteria [71]**:** formerly called blue-green algae, the cyanobacteria are photosynthetic bacteria commonly found in marine systems.

cyclonic gyres [99]**:** regions of large water movement (anticlockwise in northern hemisphere and clockwise in southern hemisphere). They result in water being upwelled from below the thermocline into surface waters as the **Coriolis effect** produces slopes in the pycnocline.

decibar [16]**:** unit of pressure equal to one tenth of atmospheric pressure at sea level. Pressure in the ocean increases by approximately 1 decibar per metre increase in depth.

density [10]**:** the mass of a sample of seawater divided by the volume of the sample. Density is expressed in units of $kg\,m^{-3}$. The density of seawater is typically a little over $1000\,kg\,m^{-3}$. The extra density above $1000\,kg\,m^{-3}$ is called sigma-t. For example, a seawater of density $1020\,kg\,m^{-3}$ has a sigma-t value of $20\,kg\,m^{-3}$.

density current [14]**:** a current that is produced by differences in **density** of seawater, with the denser water flowing underneath the less dense.

diapause [77]**:** a state of dormancy.

diffuse attenuation coefficient [52]**:** (symbol k in equation 7.2) controls the rate at which natural sunlight decays with depth in the sea. k has units of m^{-1}. At a depth equal to $1/k$, solar irradiance has decreased to $1/e$ (about 37%) of its surface value.

dinoflagellates [71]**:** a group of unicellular organisms that have two flagella which they use to move. They can be 'naked' or 'armoured'. The latter are covered in plate-like structures made of cellulose. Many dinoflagellate species are photosynthetic, but other dinoflagellates capture food, which they digest. Some species have both photosynthetic and non-photosynthetic modes of obtaining energy (*mixotrophs*).

diurnal tide [38]**:** a tide with just one high water per day.

entrainment [58]**:** in the context of turbulent layers in the sea, this is the process by which one turbulent layer can expand into another, less turbulent, layer. A good example is the deepening of the surface wind mixed layer after a storm.

epipelagic [67]**:** the zone in the ocean ranging from the surface down to about 200 m water depth.

eukaryote [67]**:** a cell or an organism comprised of cells that have a membrane-enclosed nucleus and usually other organelles (cf. **prokaryote**).

Eulerian observations [28, 133]**:** these are observations made at a fixed point in the sea. Changes in, for example, Eulerian observations of temperature, may be produced by local heating or cooling, or they may be produced by a current bringing warmer water to the observing site (*see also* Lagrangian observations).

eutrophication [126]**:** a process by which increased growth of primary producers (photosynthetic algae and bacteria) is induced by a change in the physical or chemical environment. Normally it is caused by increased nutrient concentrations in the water.

fetch [20]**:** the distance over which the wind acts upon the sea surface. Fetch is an important parameter controlling **wave heights**.

fix [89]**:** this term is often used to describe the assimilation of carbon by photosynthetic organisms, i.e. taking inorganic carbon and converting it to organically bound carbon.

flagellum [71]**:** (plural *flagella*). A filamentous organ of motility used by many groups of bacteria, protozoans and algae. An organism with a flagellum or flagella is sometimes described as being *flagellated*.

foraminifera [72]**:** commonly called *Forams*. A group of protozoans commonly found in the marine plankton. There are other formainifer species that are benthic. The tests of foraminifers are made of calcium carbonate and these can sink from the pelagic and accumulate on the sea floor in such concentrations they are referred to as *foraminiferal oozes*.

geostrophic balance [32]**:** this is the name given to the situation in which a pressure force acting on

a parcel of water is exactly balanced by the Coriolis force arising from the Earth's spin. A geostrophic balance is common, at least approximately, in many oceanic flows.

grease ice [113]**:** the earliest stage of sea ice formation, when ice crystals rise through the water column to accumulate in slicks of ice that are strongly influenced by wind and surface water movement.

group velocity [23]**:** the speed at which a group of waves with a range of wavelengths travels.

growth yield [102]**:** the efficiency by which the energy taken in is converted into growth is called the growth yield, and is normally somewhere between 10 and 30%.

hadalpelagic [67]**:** the deepest waters in the ocean, below 6000 m water depth.

halocline [14]**:** a region in the ocean where salinity changes with depth.

haloplankton [69]**:** organisms that live in the plankton for the whole of their life. (cf. **meroplankton**).

ice pancakes [114]**:** early stage of ice formed in turbulent water conditions. Predominantly circular pieces of ice form about 3 cm to 3 m in diameter, generally less than 30 cm in thickness and with raised edges.

internal waves [14]**:** waves that form in the body of the ocean on an interface between water of different densities.

intertidal [67]**:** the region of a coastal shoreline that is periodically exposed to the air at low tides.

irradiance [56]**:** light energy falling on unit area of a horizontal surface in unit time, U.

Kelvin wave [39]**:** progressive waves affected by Earth rotation. In the northern hemisphere, the crest of the wave slopes up to the right (and the trough of the wave down to the right) looking in the direction of wave travel. A Kelvin wave travelling down a channel therefore has greater amplitude on the right-hand shore.

kinematic viscosity [79]**:** in everyday terms, the viscosity of a liquid is a measure of its stickiness. Kinematic viscosity is a particular measure of this stickiness that includes the fluid density. It has units of $m^2 s^{-1}$. The kinematic viscosity of water at 20 ˚C is about 10^{-6} $m^2 s^{-1}$.

Lagrangian observations [28]**:** these are observations made following the flow, for example from a drifting buoy. E.g. changes in Lagrangian observations of temperature must be produced by local heating or cooling (cf. **Eulerian observations**).

Lambert-Beer Law [60]**:** states that **irradiance** decreases exponentially with depth in the sea.

lock gate experiment [5]**:** a laboratory experiment in which waters of two different densities are separated by a lock gate. When the lock gate is removed, a **density current** is formed.

lunar hour [39]**:** one twelfth of a semi-diurnal tidal cycle, equal to one hour and two minutes.

meroplankton [69]**:** organisms that have part of their life history in the plankton and other stages on the sea floor or intertidal shores.

mesopelagic [67]**:** the zone in the ocean ranging from about 200 m down to 1000 m water depth.

microbial loop [104]**:** this is the pathway where organic matter (both dissolved and particulate) produced by all organisms in the whole food web is broken down by bacteria. Organic matter derived from terrestrial sources is also broken down in a similar manner. A consequence of the microbial loop is the regeneration of inorganic nutrients for new phytoplankton growth.

neap tides [38]**:** a period with relatively low tidal range that occurs twice each month.

nekton [67]**:** those animals that are strong enough swimmers to be able to move against a water current (e.g. large crustaceans, fish, squid and whales) (cf. **plankton**).

nodal line [42]**:** a line in a standing wave along which there is no vertical motion of the water surface. The first nodal line forms ¼ wavelength from the reflecting shore, the second ¾ wavelength, and so on. Although there is no vertical motion at the nodal line, the horizontal currents are faster than at other places in the wave.

orbitals [20]**:** the name given to the motion of water parcels beneath waves. For waves in deep water, the orbitals are circular and diminish in diameter with increasing depth. As waves move into shallower water, the orbitals become first elliptical and then flattened, just a backwards and forwards motion (called rectilinear motion).

pelagic [67]**:** this is the open sea and includes the whole water column. Pelagic organisms are those that live in the open waters, whereas benthic organisms live on the benthos.

phase speed (of waves) [20]**:** the speed at which the crest of a wave of a given wavelength travels.

photic zone [00]**:** the region of the ocean close to the surface where there is sufficient light for photosynthesis.

photosynthetically active radiation [56]**:** (or PAR) is the name given to the portion of the sun's energy that can be used in photosynthesis. It spans the wavelength range 400–700 nanometres. This is also approximately the range over which the human eye is sensitive and so PAR is equivalent to 'visible light'.

Phytoplankton [68]**:** photosynthetic organisms of the **plankton**.

plankton [67]**:** Pelagic organisms (many microscopic) that drift or float passively with the prevailing currents in an ocean, sea or lake. Plankton include many microscopic organisms, such as bacteria, algae, protozoans, animal larvae and larger animals that cannot swim against water currents. It is divided into **zooplankton**, **phytoplankton** and **bacterioplankton.**

potential energy [49]**:** the energy an object possesses by virtue of its height in the earth's gravitational field. Potential energy can be calculated as mgh where m is the mass of an object, g is the acceleration due to gravity and h is the height of the object above or below a fixed reference. It therefore takes energy to lift an object (and increases its potential energy) and this energy is released when the object falls.

pressure [16]**:** force per unit area. The pressure in the ocean increases by approximately 1 **decibar** for every metre increase in depth.

pressure differences [32]**:** along a horizontal surface, are responsible for driving many oceanic flows. If the sea surface is sloping, for example, this produces pressure differences acting to produce a flow in the direction of the downwards slope. Pressure differences can also be caused by differences in water density.

prokaryote [67]**:** a cell or organism lacking a nucleus and other membrane-bound organelles.

pseudopod [71]**:** a pseudopod means 'fake foot' and refers to a temporary extension from a cell or unicellular organisms. Pseudopods can be used for catching prey, and for motility. In Foraminifers the pseudopodia are called *reticulopodia*, and these can form net-like structures to catch food items.

pycnocline [14]**:** a region in the ocean where density is changing with depth. To maintain vertical stability, the density must increase with increasing depth, otherwise, overturning will occur.

reduced gravity [11]**:** an object placed in water experiences buoyancy (*see* **Archimedes' Principle**) and therefore weighs less than it does in air. This concept is expressed in terms of the reduced gravity, which is given the symbol g'.

refraction [23]**:** refraction of waves refers to a change in direction of the wave travel associated with a change in their speed. The speed of waves in shallow water, for example, depends on water depth. A part of a wave crest in shallow water will therefore travel more slowly than a part of the same crest in deeper water. This produces a bending of the wave crest and a change in the direction of wave travel (*see also* **Snell's Law**).

resonance [43]**:** a body of water contained within shores has a natural period at which it will oscillate if set in motion. If a regular force is applied at this same period, large oscillations result, called resonance.

Reynolds number [80]**:** the product of the flow velocity and flow length scale divided by the kinematic viscosity of water. Turbulence becomes important at high Reynolds numbers and can be neglected compared to viscous forces (associated with the stickiness of the water) at low Reynolds numbers. In practice, turbulence is always important at scales above microscopic, but viscous forces dominate the motion of microscopic animals.

salinity [1]**:** the salt content of seawater. Salinity is properly expressed in Practical Salinity Units (PSU), which is a ratio and therefore dimensionless. In most cases, the salinity in PSU corresponds closely to the number of grams of salt in 1 kg of water (or parts per thousand).

salt fingers [17]**:** when warm salt water is placed above cold fresh water in a laboratory tank, rising 'fingers' of salt water can be observed to form on the interface. As the fingers rise, they gain heat from the surrounding water faster than they lose salt. This produces a reduction in density and an increase in the buoyancy of the salt finger, which therefore continues to rise.

Scombrids [104]**:** a group of fish including the mackerel and tuna. They are generally fast swimmers and include many of the commercially valuable fish species.

'sea' [19]**:** the name given to relatively short **wavelength** waves which have been produced locally by the wind.

sea ice biota [112]**:** organisms that are found living within or closely associated with sea ice for some, or part, of their seasonal life histories. Sometimes referred to as 'sympagic biota'.

seasonal thermocline [47]**:** the boundary between the sun-warmed surface mixed layer and deeper, cooler water that forms in seas in temperate latitudes in the spring and lasts through the summer and early autumn. In late autumn and winter, the seasonal thermocline is destroyed by a combination of surface cooling and strong winds.

Secchi disc [61]**:** a (usually) white disc that is lowered on a marked rope into the sea to give a simple measure of water transparency. The depth at which the disc is no longer visible is called the Secchi depth.

semi-diurnal tide [36]**:** a tide with two high waters per day.

Snell's Law [25]**:** is a law adapted from optics that describes the effect of **refraction** on the angle wave crests make with the shore. According to Snell's Law, the sine of the angle the wave crests make with the shore divided by the wave speed remains constant. As the waves move into shallower water near the shore, their speed is reduced and so, according to this law, the crests become more parallel to the shore.

spring tide [38]**:** a period with relatively large tidal range, which occurs twice per month.

standing wave [42]**:** is a wave motion created by the interference of two equal progressive waves travelling in opposite directions. A standing wave will have one or more nodes where the vertical motion in the two waves cancels out.

Stokes' Law [81]**:** governs the falling speed of objects in a viscous fluid. Objects sinking in the ocean quickly reach a terminal velocity in which their weight is balanced by the frictional force arising from their motion through water. The sinking speed increases with the square of the diameter of the object.

stratification [14]**:** the vertical layering of water of different density.

subtidal [67]**:** this is the area extending below the lowest low-tide point, and so is permanently submerged by water.

surface mixed layer [47]**:** is the name of the surface layer of the sea, warmed by the sun and stirred by the wind.

Sverdrup [28]**:** a unit for the volume of water transported by ocean currents, equal to one million cubic metres per second.

swell [19]**:** long **wavelength**, regular waves, with discernable long crests.

temperature–salinity (or T/S) diagram [6]**:** a graph on which the temperature of seawater at a given depth is plotted against the salinity at the same depth. T/S diagrams are useful for identifying water masses and their mixing in the ocean.

temperature staircase [17]**:** a series of 'steps' on a plot of water of water temperature against depth which has been observed in places in the ocean where it is thought that the difference in the rate of diffusion of salt and heat might be important. The steps are thought to form as a result of **salt fingering**.

test [72]**:** this is the shell-like structures produced by foraminifers. They are made up of chambers and constructed from calcium carbonate.

thermocline [5]**:** is the name given to a region in which temperature in the ocean is changing with depth. The main ocean thermocline separates sun-warmed surface water from water that is close to freezing near the ocean floor.

thermohaline circulation [5]**:** the relatively slow movement of the water in the deep ocean produced by differences in **density** of the water. The density differences, in turn, are produced by variations of temperature and **salinity**.

tidal mixing fronts [52]**:** tidal mixing fronts, found in shelf seas in temperate latitudes in the summer, separate thermally stratified water from vertically mixed water.

tidal range [37]**:** the vertical distance between low water and high water during a single tidal cycle.

tide generating force [35]**:** the tide generating force of the Moon on the Earth is the difference between the Moon's gravitational force at a point on the Earth's surface and the Moon's gravitational force at the Earth's centre.

transfer efficiency [102]**:** the efficiency (energy

gain–energy loss) by which energy is passed between one trophic level and another in a food chain or web.

trophosome [116]**:** an organ in an animal that contains high densities of symbiotic bacteria which produce organic matter that the animal can use; e.g. in Riftia spp. found in hydrothermal vent systems.

tsunami [19]: the name given to a wave produced by an underwater disturbance, such as a submarine landslide or earthquake. Tsunamis have long wavelengths and so their speed, even in the deep ocean, is given by the shallow-water wave speed equation. Their wave height in the deep ocean is quite small, but grows dramatically as the tsunami moves into shallow water near a coast.

turbidity current [15]**:** an example of a density current in which the differences in density are produced by suspended particles.

volume transport [28]**:** the volume of water transported by an ocean current in unit time (*see also* **Sverdrup**).

wave dispersion [22]**:** the result of the speed of waves in deep water, depending on their period. Long period waves therefore outstrip shorter period waves, and waves of different period travelling away from a storm become separated.

wave height [20]**:** the vertical distance between the **crest** and the **trough** of a wave.

wave length [20]**:** the horizontal distance between one wave crest and the next.

wave period [20]**:** the time that elapses between one wave crest and the next passing a fixed point.

western intensification [27]**:** the name given to the formation of narrow, fast-moving currents on the western margins of the oceans, compared to the weaker, broader currents on the eastern margins. The effect is produced by the need to conserve angular momentum.

wind waves [19]**:** waves produced by the action of the wind on the sea.

zooplankton [69]**:** animals and protozoans in the plankton.

Further Reading

There are many excellent and comprehensive books on all aspects of oceanography. Some concentrate more on the physics, and others more on the biology and/or chemistry. Below is a list that we think gives a good representation of the literature to follow on from some of the basic concepts and ideas presented in this short introduction. One of the best web resources can be found in *Oceanus* published by the Woods Hole Oceanographic Institute. (http://www.whoi.edu/oceanus/index.do).

Bigg, G. (2003) *The Oceans and Climate*. Cambridge University Press.

Cockell, C., Corfield, R., Edwards, N. & Harris, N. (2007) *An Introduction to the Earth–Life System*. Cambridge University Press.

Denny, M. (2008) *How the Ocean Works*. Princeton University Press.

Emerson, S.R. & Hedges, J.I. (2008) *Chemical Oceanography and the Marine Carbon Cycle*. Cambridge University Press.

Falkowski, P.G. & Raven, J.A. (2007) *Aquatic Photosynthesis*. (2nd Edition) Princeton University Press.

Kaiser, M.J, Attrill, M., Jennings, S., Thomas, D.N., Barnes, D., Brierley, A.S., Hiddink, J.G., Kaartokallio, H., Polunin, N. & Raffaelli, D. (2011) *Marine Ecology – Processes, Systems and Impacts*. (2nd Edition) Oxford University Press.

Kirchman, D.L. (2008) *Microbial Ecology of the Oceans*. (2nd Edition) Wiley Blackwell.

Kirk, J.T.O. (2010) *Light and Photosynthesis in Aquatic Ecosystems*. Cambridge University Press.

Lalli, C.M. & Parsons, T.R. (2004) *Biological Oceanography: An Introduction*. (2nd Edition) Butterworth-Heinemann.

Mann, K.H. & Lazier, J.R.N. (2006) *Dynamics of Marine Ecosystems: Biological–physical interactions in the oceans*. (3rd Edition) Wiley Blackwell.

Miller, C.B. (2004) *Biological Oceanography*. Wiley Blackwell.

Nybakken, J.W. & Bertness, M.D. (2005) *Marine Biology: An ecological approach*. (6th Edition) Benjamin Cummings.

Open University. (1995) *Seawater, Its Composition, Properties and Behaviour*. Butterworth-Heinemann.

Open University. (2000) *Waves, Tides and Shallow-Water Processes*. Butterworth-Heinemann.

Open University. (2005) *Marine Biogeochemical Cycles*. Butterworth-Heinemann Publishing.

Pinet, P.R. (2003) *Invitation to Oceanography*. (3rd Edition) Jones & Bartlett Publishing.

Pugh, D.T. (2004) *Changing sea levels: Effects of tides, weather and climate*. Cambridge University Press.

Sarmiento, J.L. & Gruber, N. (2006) *Ocean Biogeochemical Dynamics*. Princeton University Press.

Talley, L.D., Pickard, G.L., Emery, W.J. & Swift, J.H. (2011) *Descriptive Physical Oceanography: An Introduction*. Elsevier Science & Technology.

Thurman, H.V. & Burton, E.A. (2001) *Introductory Oceanography*. Prentice Hall.

Wells, N.C. (2012) *The Atmosphere and Ocean: A physical introduction*. (3rd Edition) Wiley Blackwell.

There are two works that have inspired us greatly, and although a bit dated; both are worth reading to get a sense of the excitement of studying oceanography:

Hardy, A. (1967) *Great waters*. Harper & Row, London, New York.

Fogg, G.E. (1991) *Tansley Review No. 30. The phytoplankton ways of life*. New Phytologist, 118, 191–232.

There are many research institutes and university departments around the world where oceanographic research is carried out. Here is a list of just a few, where useful, up-to-date information can be obtained with links to other web resources:

Alfred Wegener Institute for Polar and Marine Research, Germany – http://www.awi.de/en/home

Australian Institute of Marine Science – http://www.aims.gov.au/

Bigelow Laboratory for Ocean Sciences, USA – http://www.bigelow.org/

Centre for Environment, Fisheries & Aquaculture Science, UK – http://www.cefas.defra.gov.uk/

French Research Institute for Exploration of the Sea – http://wwz.ifremer.fr/institut_eng/

GEOMAR, Germany – http://www.geomar.de/en/

Institute of Marine Research, Norway – http://www.imr.no/en

Marine Biological Laboratory, USA - http://www.mbl.edu/

Max Plank Institute for Marine Microbiology, Germany – http://www.mpi-bremen.de/en/

Monterey Bay Aquarium Research Institute, USA – http://www.mbari.org/

National Aeronautics and Space Admisistration (NASA) Oceanography, USA – http://nasascience.nasa.gov/earth-science/oceanography/

National Institute of Water and Atmospheric Research, New Zealand – http://www.niwa.co.nz/our-science/coasts-and-oceans

National Oceanography Center, UK – http://www.noc.soton.ac.uk

National Oceanic and Atmospheric Admisitration, USA – http://www.noaa.gov

National Snow and Ice Data Center, USA – http://nsidc.org/

Royal Netherlands Institute for Sea Research – http://www.nioz.nl/

SCRIPPS Institution of Oceanography, USA – http://sio.ucsd.edu/

Sir Alister Hardy Foundation for Ocean Science, UK – http://www.sahfos.ac.uk/

Stazione Zoologica Anton Dohrn, Italy - http://www.szn.it/

Woods Hole Oceanographic Institute, USA – http://www.whoi.edu

Sourced illustrations

The following individuals/organisations have very kindly supplied images (figure numbers in parenthesis), and in each instance they hold the original copyright:

Alice Alldredge, University of California, Santa Barbara (11.9), Argo Program (http://www.argo.ucsd.edu) (4.3), Ingo Arndt, Ingo Arndt Wildlife Photography (8.7A, 8.10–8.14), Riitta Autio (8.4), Elanor Bell (8.7B), David G. Bowers (Figures 1.2, 1.4–1.9, 2.1, 2.4–2.5, 2.7, 3.1, 3.4–3.6, 4.1, 4.6–4.8, 5.1–5.10; 6.1–6.5; 7.2–7.4, 7.5A, 7.6, 7.7B, 7.9, 7.11, 14.3–14.6), Mark Brandon (14.10B), Andrew Brierley (8.15, 14.10A), Dudley Chelton (Fig.i.i), Finlo Cottier (8.16A), Gerhard Dieckmann (8.17, 14.12, 14.15C), Daniel Eriksson/Ålandstidningen (12.9), Luis Gimenez (8.3), Päivi Hakanen (8.5, 8.8B, 10.1B), Karen Haywood (14.14), Kevin Horsburgh (6.8), Cecil Jones (6.7A), Andrew Juhl and Christopher Krembs (9.3), Christopher Krembs (12.6), Amy Leventer (14.13), Ian Lucas (8.18), Jonathon Malarkey (5.6–7), National Oceanic and Atmospheric Administration (4.4, 9.1, 9.6, 9.8, 9.9, 11.3, 12.8, 13.2, 14.8, 14.9), National Snow and Ice Data Center, Boulder, Colorado, USA (13.3, 13.4), Anna Pienkowski (8.6), Stefan Rahmstorf (4.2), David Roberts (2.3, 2.6, 3.3, 10.1A, 11.2, 14.1B,C), SeaWiFS Project/NASA/Goddard Space Flight Center/GeoEye (8.9, 13.9A), SeaWiFS Project/NASA/Goddard Space Flight Center/ORBIMAGE (7.10, 10.5, 10.6), Mariano Sironi (11.1), Craig Smith, University of Hawaii (11.11), Howard Spero (8.7C), Jacquelin Stefels

(8.8A), Reto Stockli, NASA Earth Observatory (1.1), David N. Thomas (2.2, 3.2, 6.7B, 7.1, 7.5B, 7.7A, 7.11, 8.1, 8.2, 8.16B,C, 9.2, 9.5, 9.7; 10.2–10.4; 10.7, 11.5–11.8; 11.10, 12.1–12.5; 13.1, 13.5, 13.7, 13.8, 13.9B, 14.1A, 14.1D, 14.2, 14.11, 14.15A,B, 14.16, 14.17), Naomi J. Thomas (8.17 C,D, 9.4), Woods Hole Oceanographic Institution (12.7).

Satellite data were received and processed by the NERC Earth Observation Data Acquisition and Analysis Service (NEODAAS) at Dundee University and Plymouth Marine Laboratory (www.neodaas.ac.uk). SeaWiFS data courtesy of the NASA SeaWiFS project and Orbital Sciences (6.6, 14.7).

Figure 4.5 was adapted from Sannino, G., Bargagli, A. and Artale, V. (2002) *Numerical modelling of the mean exchange through the strait of Gibraltar*. Journal of Geophysical Research, 107 (C8), 3094, doi 10.1029/2001JC000929.

Figure 7.8 was adapted from Ní Rathaille, A. (2007) PhD thesis, National University of Ireland, Galway.

Figure 11.4 was taken from Link, J. (2002) *Does food web theory work in marine ecosystems?* Marine Ecology Progress Series, 230, 1–9.

Figure 13.6 was adapted from Wolf-Gladrow, D.A., Riebesell, U., Burkhardt, S. and Bijma, J. (1999) *Direct effects of CO_2 concentration on growth and isotopic composition of marine plankton*. Tellus B, 51, 461–476.

We are grateful to Sandra Mather for redrawing many of the images.